Advanced Processing and Manufacturing Technologies for Structural and Multifunctional Materials VII

Advanced Processing and Manufacturing Technologies for Structural and Multifunctional Materials VII

A Collection of Papers Presented at the 37th International Conference on Advanced Ceramics and Composites January 27–February 1, 2013 Daytona Beach, Florida

Edited by
Tatsuki Ohji
Mrityunjay Singh

Volume Editors
Soshu Kirihara
Sujanto Widjaja

The American Ceramic Society

WILEY

Library of Congress Cataloging-in-Publication Data is available.

ISBN: 978-1-118-80773-6
ISSN: 0196-6219

Printed in the United States of America.

10 9 8 7 6 5 4 3 2 1

Contents

*Presented at the ICACC 2012 Conference

Preface

The 7th International Symposium on Advanced Processing and Manufacturing Technologies for Structural and Multifunctional Materials and Systems (APMT) was held during the 37th International Conference on Advanced Ceramics and Composites, in Daytona Beach, FL, January 27–February 1, 2013. The aim of this international symposium was to discuss global advances in the research and development of advanced processing and manufacturing technologies for a wide variety of non-oxide and oxide based structural ceramics, particulate and fiber reinforced composites, and multifunctional materials. A total of 56 papers, including invited talks, oral presentations, and posters, were presented from 15 countries (USA, Japan, Germany, China, Korea, UK, Switzerland, Canada, Estonia, India, Italy, Luxembourg, Serbia, and Malaysia). The speakers represented universities, industry, and research laboratories.

This issue contains 19 invited and contributed papers including 3 presented in Engineering Ceramics Summit of the Americas; all were peer reviewed according to The American Ceramic Society review process. The latest developments in processing and manufacturing technologies are covered, including green manufacturing, smart processing, advanced composite manufacturing, rapid processing, joining, machining, and net shape forming technologies. These papers discuss the most important aspects necessary for understanding and further development of processing and manufacturing of ceramic materials and systems.

The editors wish to extend their gratitude and appreciation to all the authors for their cooperation and contributions, to all the participants and session chairs for their time and efforts, and to all the reviewers for their valuable comments and suggestions. Financial support from the Engineering Ceramic Division and The American Ceramic Society is gratefully acknowledged. Thanks are due to the staff of the meetings and publication departments of The American Ceramic Society for their invaluable assistance.

We hope that this issue will serve as a useful reference for the researchers and technologists working in the field of interested in processing and manufacturing of ceramic materials and systems.

TATSUKI OHJI, *Nagoya, Japan*
MRITYUNJAY SINGH, *Cleveland, USA*

Introduction

This issue of the Ceramic Engineering and Science Proceedings (CESP) is one of nine issues that has been published based on manuscripts submitted and approved for the proceedings of the 37th International Conference on Advanced Ceramics and Composites (ICACC), held January 27–February 1, 2013 in Daytona Beach, Florida. ICACC is the most prominent international meeting in the area of advanced structural, functional, and nanoscopic ceramics, composites, and other emerging ceramic materials and technologies. This prestigious conference has been organized by The American Ceramic Society's (ACerS) Engineering Ceramics Division (ECD) since 1977.

The 37th ICACC hosted more than 1,000 attendees from 40 countries and approximately 800 presentations. The topics ranged from ceramic nanomaterials to structural reliability of ceramic components which demonstrated the linkage between materials science developments at the atomic level and macro level structural applications. Papers addressed material, model, and component development and investigated the interrelations between the processing, properties, and microstructure of ceramic materials.

The conference was organized into the following 19 symposia and sessions:

Symposium 1	Mechanical Behavior and Performance of Ceramics and Composites
Symposium 2	Advanced Ceramic Coatings for Structural, Environmental, and Functional Applications
Symposium 3	10th International Symposium on Solid Oxide Fuel Cells (SOFC): Materials, Science, and Technology
Symposium 4	Armor Ceramics
Symposium 5	Next Generation Bioceramics
Symposium 6	International Symposium on Ceramics for Electric Energy Generation, Storage, and Distribution
Symposium 7	7th International Symposium on Nanostructured Materials and Nanocomposites: Development and Applications

Symposium 8	7th International Symposium on Advanced Processing & Manufacturing Technologies for Structural & Multifunctional Materials and Systems (APMT)
Symposium 9	Porous Ceramics: Novel Developments and Applications
Symposium 10	Virtual Materials (Computational) Design and Ceramic Genome
Symposium 11	Next Generation Technologies for Innovative Surface Coatings
Symposium 12	Materials for Extreme Environments: Ultrahigh Temperature Ceramics (UHTCs) and Nanolaminated Ternary Carbides and Nitrides (MAX Phases)
Symposium 13	Advanced Ceramics and Composites for Sustainable Nuclear Energy and Fusion Energy
Focused Session 1	Geopolymers and Chemically Bonded Ceramics
Focused Session 2	Thermal Management Materials and Technologies
Focused Session 3	Nanomaterials for Sensing Applications
Focused Session 4	Advanced Ceramic Materials and Processing for Photonics and Energy
Special Session	Engineering Ceramics Summit of the Americas
Special Session	2nd Global Young Investigators Forum

The proceedings papers from this conference are published in the below nine issues of the 2013 CESP; Volume 34, Issues 2–10:

- Mechanical Properties and Performance of Engineering Ceramics and Composites VIII, CESP Volume 34, Issue 2 (includes papers from Symposium 1)
- Advanced Ceramic Coatings and Materials for Extreme Environments III, Volume 34, Issue 3 (includes papers from Symposia 2 and 11)
- Advances in Solid Oxide Fuel Cells IX, CESP Volume 34, Issue 4 (includes papers from Symposium 3)
- Advances in Ceramic Armor IX, CESP Volume 34, Issue 5 (includes papers from Symposium 4)
- Advances in Bioceramics and Porous Ceramics VI, CESP Volume 34, Issue 6 (includes papers from Symposia 5 and 9)
- Nanostructured Materials and Nanotechnology VII, CESP Volume 34, Issue 7 (includes papers from Symposium 7 and FS3)
- Advanced Processing and Manufacturing Technologies for Structural and Multi functional Materials VII, CESP Volume 34, Issue 8 (includes papers from Symposium 8)
- Ceramic Materials for Energy Applications III, CESP Volume 34, Issue 9 (includes papers from Symposia 6, 13, and FS4)
- Developments in Strategic Materials and Computational Design IV, CESP Volume 34, Issue 10 (includes papers from Symposium 10 and 12 and from Focused Sessions 1 and 2)

The organization of the Daytona Beach meeting and the publication of these proceedings were possible thanks to the professional staff of ACerS and the tireless dedication of many ECD members. We would especially like to express our sincere thanks to the symposia organizers, session chairs, presenters and conference attendees, for their efforts and enthusiastic participation in the vibrant and cutting-edge conference.

ACerS and the ECD invite you to attend the 38th International Conference on Advanced Ceramics and Composites (http://www.ceramics.org/daytona2014) January 26-31, 2014 in Daytona Beach, Florida.

To purchase additional CESP issues as well as other ceramic publications, visit the ACerS-Wiley Publications home page at www.wiley.com/go/ceramics.

SOSHU KIRIHARA, *Osaka University, Japan*
SUJANTO WIDJAJA, *Corning Incorporated, USA*

Volume Editors
August 2013

CREATION OF SURFACE GEOMETRIC STRUCTURES BY THERMAL MICRO-LINES PATTERNING TECHNIQUES

Soshu Kirihara, Satoko Tasaki
Joining and Welding Research Institute
Osaka University
11-1 Mihogaoka Ibaraki, Osaka 567-0047, Japan

Yusuke Itakura
Graduate School of Engineering
Osaka University
2-1 Yamadaoka Suita, Osaka 565-0871, Japan

ABSTRACT

Thermal micro lines patterning techniques were newly developed as novel technologies to create geometrical intermetallics patterns for mechanical properties modulations of metal substrates. Pure copper particles were dispersed into the photo solidified liquid resins, and these slurries were spread on aluminum substrates. Micro patterns with fractal structures of Hilbert curves and dendritic lines were drawn and fixed by an ultra violet laser scanning. The formed patterns on the substrates were heated in an argon atmosphere, and the intermetallic or alloy phases with high hardness were created through reaction diffusions. The mechanical properties and surface stress distributions were measured and simulated by a tensile stress test and finite element method.

INTRODUCTION

Fractal geometries with self-similarity can be applied to modulate various flows in engineering fields [1,2]. Geometric networks with the fractal structure of intermetallic compounds patterned on light metals can strengthen whole materials efficiently by controlling surface stress distributions intentionally. In our research group, three dimensional metal and ceramics lattices with dendritic structures have been created and inserted into various matrices successfully to control stress and heat distributions [3,4]. Considering the next generation, mechanical properties enhancements by novel surface treatments will be expected to contribute novel materials processing of rare metals free. In this investigation, micro patterns composed of copper aluminide had been created on pure aluminum substrates by using a laser scanning stereolithography and a reaction diffusion joining. Microstructures and composite distributions in the vicinity of formed alloy and metal interfaces were observed and analyzed by using an electron microscope. Load dispersion abilities of the network were evaluated by using conventional mechanical tests and compared with simulated and visualized profiles by using a numerical analysis simulation.

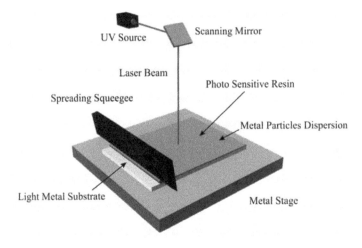

Figure 1.   Schematically illustrated system configurations of a laser scanning stereolithography.

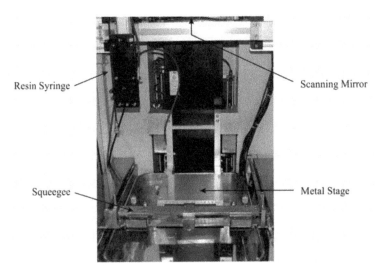

Figure 2.   An appearance photograph of the used laser scanning stereolithography equipment.

EXPERIMENTAL PROCEDURE

Self-similar patterns of Hilbert curve with stage numbers 1, 2, 3 and 4 of the fractal line structures were designed by using a computer graphic application, and these graphic images were converted into the numerical data sets by computer software. These patterns of 25×25 mm in whole size were composed of arranged lines of 400 μm in width. These graphic models were transferred into the processing apparatus as operating data sets. Metal particles were patterned on a metal substrate by using a stereolithographic system. The pure copper particles of 50 μm in average diameter dispersed into a photo sensitive urethane resin at 60 volume percent. The mixed resin paste was spread with 100 μm in layer thickness on an aluminum substrate of 30×30×2 mm in size by using a mechanically moved knife edge as shown in Fig. 1. An ultraviolet laser beam of 355 nm in wavelength and 100 μm in beam spot was scanned on the resin surface according to the computer operation. A two dimensional solid pattern was obtained by a light induced photo polymerization. Figure 2 shows the appearance of the stereolithographic system. Subsequently, the dendiritic fractal patterns with the self-similarity of stage number 3 were adopted as the geometrical patterns. The patterns of 20×80 mm in whole size were composed of arranged lines of 400 and 8000 μm in width and length. The mixed resin paste with the pure copper particles was patterned on the aluminum specimen of 20×80×2 mm in size of the parallel part by using the stereolithography. After removing uncured resin by ultrasonic cleaning in ethanol solvent, the sample was heated to dewax the resin and create a self-similar pattern composed of copper aluminides at 600 °C above eutectic temperature for 4 hs of holding time in an argon atmosphere. Microstructures and composite distributions were observed by a scanning electron microscopy (SEM) and an energy dispersive X-ray spectroscopy (EDS), respectively. Stress distribution in the patterned material during uniaxial tension tests was simulated by a finite element method (FEM) calculation.

RESULTS AND DISCUSSION

The copper aluminide micro pattern with the fractal structure of Hilbert curve was formed successfully on the substrate. Figure 3 shows the formed fractal polyline of the number 3 in fractal stage. The micrometer order geometric structure was composed of fine intermetallics lines of 450 μm in width. The part accuracy of these microline patterns was estimated as 10 % approximately. The copper aluminide composite was formed widely comparing with the designed line width, though the reaction diffusion between the copper and aluminum. The EDS measurement results suggested that copper concentrated in the micro network and formed intermetallic phase of $CuAl_2$. Microscopic defects were not found in the formed intermetallics layers though the SEM observations. The smooth interfaces were obtained between copper aluminide and the aluminum substrate. During the heat treatment at the high temperature, eutectic reaction with liquid phase formation occurred between the pure copper particles and the pure aluminum substrate. After the solidification of molten alloys, the dual phase microstructure of the intermetallics and alloys composites can exhibit the higher mechanical

Figure 3.   A copper aluminide Hilbert curve patterned on an aluminum substrate.

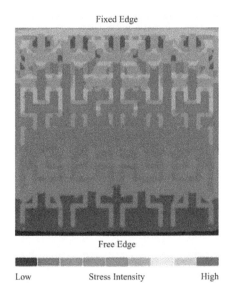

Figure 4.   A stress distribution on the fractal pattern of intermetallics visualized by using FEM.

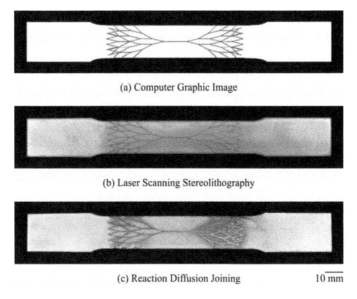

(a) Computer Graphic Image

(b) Laser Scanning Stereolithography

(c) Reaction Diffusion Joining   10 mm

Figure 5.   The dendrite patterns formed by the stereolithography and reaction diffusion.

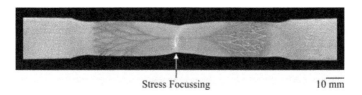

Stress Focussing   10 mm

Figure 6.   A tensile specimen with the dendritic fractal pattern after mechanical test.

Low ▬▬▬▬▬▬▬▬▬▬▬ High

Figure 7.   A simulated and visualized stress distribution on the patterned test specimen.

strength. The stress distributions on the patterned surfaces were visualized for the Hilbert curve of stage number 3 through the numerical simulation as shown in Fig. 4. The required mechanical properties of Young's modulus were defined along the compositional analysis and the phase identifications. The stress intensities concentrate into the vicinity of fixed edge and are distributed along the patterned lines and the corners with the higher hardness.

According to the designed dendritic model as shown in Fig. 5-(a), the real fractal pattern composed of the pure cupper dispersed urethane resin were created clearly as shown in Fig. 5-(b) by using the laser scanning stereolithography. The formed line width was measured as 400 μm. The part accuracies of geometric patterns were verified within 5 %. Through the heat treatments, the copper aluminide networks with the self similar patterns were formed by using the reaction diffusion joining as shown in Fig 5-(c). The mechanical properties of the patterned sample were measured through the tensile test as shown in Fig. 6. A small fracture crevasse is formed in the central position of the specimen perpendicularly to the intermetallics line connecting with two dendritic patterns. The tensile stresses were considered to be focused effectively by the both dendritic patterns and concentrated into the center connecting line. Through the dynamically monitoring for this artificial fracture source, real time materials life estimations will be realized effectively. The stress distributions on the patterned surface were simulated and visualized for the dendritic lines of stage number 3 through the FEM calculation as shown in Fig. 7. In the calculation process, the tensile strengths were loaded for the both edge of the specimen model. The red and blue colored areas indicate the higher and lower intensities of the bending stress, respectively. The stress intensities are concentrated into the vicinity of the center region and distributed along the patterned lines. The crossing points of the intermetalics lines with the higher hardness show the stress concentrations. The simulated and visualized results have good agreements with the measured results as shown in Fig. 6. The self-similar patterns can include the more numbers of sides and nodes in the limited area comparing with the periodic arrangements of the geometric polygon figures. Therefore, the dendritic fractal line patterns are considered to be able to disperse the mechanical stresses intentionally on the surface areas of the substrates.

CONCLUSIONS

Geometric networks with fractal structures of Hilbert curves and dendritic lines were created to modulate mechanical properties intentionally through computer aided designing and manufacturing. Microlines composed of urethane resin with copper particles dispersion were patterned successfully on pure aluminum substrates by using a stereolithography. The patterned samples were heated in an argon atmosphere to create intermetallics lines of copper aluminides through reaction diffusion between the copper and aluminum. Cracks and pores were not observed in welded interfaces by using a scanning electron microscopy. Mechanical stresses distributions along the formed intermetallics networks with were visualized and observed by numerical simulations and tensile tests. Investigated patterning techniques are considered to be efficient strengthening processes for light metals and alloys.

REFERENCES

[1] Y. Kuan, J. Chang, S. M. Lee, S. R. Lee, Characterization of a Direct Methanol Fuel Cell Using Hilbert Curve Fractal Current Collectors, *J. Power Source*, **187**, 112-122 (2009).

[2] J. Yang, H. Bin, X. Zhang, Z. Liu, Fractal Scanning Path Generation and Control System for Selective Laser Sintering, *Int. J. Mach. Tools Manuf.*, **43**, 293-300 (2003).

[3] S. Kirihara, Development of Photonic and Thermodynamic Crystals Conforming to Sustainability Conscious Materials Tectonics, *WTI Trans. Ecolo. Env.*, **154**, 103-114 (2011).

[4] Y. Uehara, S. Tasaki, S. Kirihara, Fabrication of Hard Alloys Patterns with Fractal Structures on Light Metal Substrates through Reaction Diffusion, *J. Smart Proc.*, **1**, 186-189 (2012).

MAGNETOELECTRIC PROPERTIES OF La-MODIFIED BiFeO₃ THIN FILMS ON STRONTIUM RUTHENATE (SrRuO₃) BUFFERED LAYER

Regina C. Deus[a], César R. Foschini[b], José A. Varela[c], Elson Longo[c] and Alexandre Z. Simões[a,c*]

[a] Universidade Estadual Paulista- UNESP - Faculdade de Engenharia de Guaratinguetá, Av. Dr. Ariberto Pereira da Cunha, 333, Bairro Portal das Colinas, Zip-Code: 12516-410– Guaratinguetá-SP, Brazil, Phone +55 12 3123 2228.
[b] Universidade Estadual Paulista, UNESP, Faculdade de Engenharia de Bauru, Dept. de Eng. Mecânica, Av. Eng. Luiz Edmundo C. Coube 14-01, Zip-Code: 17033-360, Bauru, SP, Brasil, , Phone +55 12 3123 2228
[c] Universidade Estadual Paulista-UNESP- Instituto de Química - Laboratório Interdisciplinar em Cerâmica (LIEC), Rua Professor Francisco Degni s/n, Zip-Code: 14800-90-, Araraquara, SP, Brazil, Phone +55 16 3301 9828.
*e-mail- alezipo@yahoo.com

ABSTRACT

This paper focus on the magnetoelectric coupling (ME) at room temperature in lanthanum modified bismuth ferrite thin film (BLFO) deposited on SrRuO₃-buffered Pt/TiO₂/SiO₂/Si (100) substrates by the soft chemical method. BLFO film was coherently grown at a temperature of 500°C. The magnetoelectric coefficient measurement was performed to evidence magnetoelectric coupling behavior. Room temperature magnetic coercive field indicates that the film is magnetically soft. The maximum magnetoelectric coefficient in the longitudinal direction was close to 12 V/cmOe. Dielectric permittivity and dielectric loss demonstrated only slight dispersion with frequency due the less two-dimensional stress in the plane of the film. The spontaneous polarization of the film was 25 μC/cm². The film has a piezoelectric coefficient, d_{33}, equal to 85 pm/V and a weak pulse width dependence indicating intrinsic ferroelectricity. Retention measurement showed no decay of polarization while piezoelectric response was greatly improved by the conductor electrode. Polarization reversal was investigated by applying dc voltage through a conductive tip during the area scanning. We observed that various types of domain behavior such as 71° and 180° domain switchings, and pinned domain formation occurred.

INTRODUCTION

Multiferroics materials[1,2] exhibit several ferroic (or antiferroic) orders simultaneously and can be for instance ferroelectric and ferroelastic[3] or ferroelectric and ferro- or ferrimagnetic[4,5]. The concomitance of several order parameters in a given material is attractive for storage applications as it offers the possibility to store twice as much information in a given memory cell volume, thereby providing an exponential increase in storage density[6]. Ferromagnetic and ferroelectric order parameters are widely used to store binary information in MRAMs[7] and FeRAMs[8], respectively but, unfortunately, ferroelectric ferromagnets (or ferrimagnets) are very scarce and the quest for a material with both large finite polarization and magnetization at room temperature is still in progress. To reach this goal, a first goal is to obtain materials with magnetoelectric coupling. Among all known multiferroics, the only compound that satisfies these criteria is bismuth ferrite (BFO). First synthesized in the late 1950s[9], BFO was shown to be a G-type antiferromagnet with a Néel temperature of 630 K by Kiselev et al[10]. Later, Sosnowska et al. showed that the magnetic order of bulk BFO is not strictly collinear and that a cycloidal modulation with a period of   62 nm is present[11]. The magnetoelectric (ME) coefficient αME=dE/dH=$dV/(tdH)$ is the most critical indicator for the magnetoelectric coupling properties in multiferroic materials, where V is the induced magnetoelectric voltage, H is the exciting ac magnetic field, and t is the thickness of the sample used for measuring V across the laminate[12]. In previous work, we grow bismuth ferrite thin films on strontium ruthenate layer in which the conducting oxide was used as both the bottom electrode and the buffer layer[13]. We have observed an improvement in the crystal growth and electric properties of ferroelectric oxide. Wang *et al.*[14] reported the considerable enhancement in the leakage, ferro/piezoelectric, and magnetic properties of the epitaxial bismuth ferrite films with conductive strontium ruthenate electrodes. Of particular interest, the SRO electrode is promising in inducing the preferred orientation and improving the film/electrode interface of ferroelectric films. Previous studies suggested that the fatigue endurance can be effectively suppressed by La doping[15]. To our knowledge, few reports are available on the magnetoelectric coefficient dependence on dc bias magnetic field of lanthanum modified bismuth ferrite thin film (BLFO) grown in oxide electrode. In this way, the magnetic and dielectric properties of BLFO thin films crystallized by the polymeric precursor method was investigated. The main focus was to evaluate the role exerted by the SRO bottom electrode on crystal structure and electrical properties of BLFO ferroic material.

EXPERIMENTAL

The SRO and BLFO thin films were prepared using the polymeric precursor method, as described elsewhere[16]. The bottom electrode thin films were spin coated on Pt/Ti/SiO$_2$/Si (100) substrates by a commercial spinner operating at 5000 revolutions/min for 30 s (spin coater KW-4B, Chemat Technology). Each layer was pre-fired at 400°C for 2 h in a conventional oven. After the pre-firing, each layer was crystallized in a microwave furnace at 700 °C for 10 minutes. Using the same procedure, the BLFO thin films were deposited by spinning the precursor solution on the desired substrates. Through this process, we have obtained thickness values of about 150 nm (5 layers) for the bottom electrode and around 300 nm for BLFO (10 layers), reached by repeating the spin-coating and heating treatment cycles. The microwave furnace used to crystallize the SRO electrode was a simple domestic model similar to that described in literature[16]. Phase analysis of the films were performed at room temperature by X-ray diffraction (XRD) using a Bragg- Brentano diffractometer (Rigaku 2000) and CuKα radiation. A PHI-5702 multifunction X-ray photoelectron spectrometer (XPS) was used, working with an Al-Kα X-ray source of 29.35 eV passing energy. The pressure in the chamber during the experiments was about $4.5.10^{-9}$ Torr. Calibration of binding energy scale was controlled using the O1's line, which appears in the photoelectron spectra of the as grown samples.The electrical properties of Pt/BLFO/SRO/Pt/Ti/SiO$_2$/Si (100) capacitor structure were measured. The upper electrodes of Pt for the electrical measurements were prepared by evaporation through a shadow mask with a 0.2 mm^2 dot area. Dielectric and ferroelectric properties of the capacitor were measured using HP 4192A impedance/gain phase analyzer and a Radiant Technology RT6000 A in a virtual ground mode, respectively. The magnetoelectric coefficient measurements in BLFO films were attained in a dynamic lock-in technique. The dc magnetic bias field was produced by an electromagnet (Cenco Instruments J type). The time-varying dc field was achieved by a programmable dc power supply (Phillips PM2810 60 V/5 A/60 W). To measure the dc magnetic field, a Hall probe was employed. Magnetization measurements were done by using a vibrating-sample magnetometer (VSM) from Quantum Design™. The magnetoelectric signal was measured by using a lock-in amplifier (EG&G model 5210) with input resistance and capacitance of 100 MΩ and 25 pF, respectively. Piezoelectric measurements were carried out using a setup based on an atomic force microscope in a Multimode Scanning Probe Microscope with Nanoscope IV controller (Veeco FPP-100). In our experiments, piezoresponse images of the films were acquired in ambient air by applying a small ac voltage with an amplitude of 2.5 V (peak to peak) and a frequency of 10 kHz while scanning the film surface. To apply the external voltage we used a standard gold coated Si$_3$N$_4$ cantilever with a spring constant of 0.09 N/m. The probing tip, with an apex radius of about 20 nm, was in mechanical contact with the uncoated film surface during the measurements. Cantilever vibration was detected using a conventional lock-in technique.

RESULTS AND DISCUSSION

XRD Phase Analysis

Figure 1 shows the XRD pattern of BLFO deposited on SRO electrode. The polycrystalline film exhibits a pure perovskite phase. Furthermore, except for the Si (100) and Pt (111) peaks, no peaks of impure phases such as Bi$_2$Fe$_4$O$_9$ and Bi$_{46}$Fe$_2$O$_{72}$ were spotted so pure phase BiFeO$_3$ films were obtained by the soft chemical method. The polycrystalline nature of the film can be attributed to the differences in nucleation energy between the ferroelectromagnetic material and the oxide electrode. The insertion of SRO can avoid a possible interfacial reaction between platinum and bismuth which can lead to undesired electrical properties. Therefore, the substitution of metallic electrodes based on noble metals like platinum by conductive oxides is an alternative to reach better electrical properties caused by the high oxygen affinity of these electrodes[17].

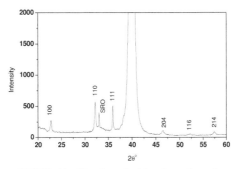

Figure 1. X-ray diffraction of BLFO thin film deposited by the polymeric precursor method on SRO electrode and annealed at 500°C in static air for 2 h.

XPS Analysis

In order to identify the chemical bonding of BFO thin films, we performed XPS studies. The spectrum expanded from 700 to 745 eV is also shown Figure 2. The 3/2 and 1/2 spin-orbit doublet components of the Fe 2p photoemission located at 711.1 and 724.6 eV, respectively were identified as Fe^{3+}. No Fe^{2+} and Fe were found. The XPS results show that BFO thin films annealed at 500°C for 2 hours has a single phase with a Fe^{3+} valence state, consistent with XRD result shown in Figure 1. The oxidation state of Fe was purely 3+, which was advantageous for producing BFO film with low leakage[18].

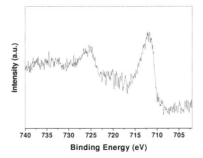

Figure 2. XPS analysis of BLFO thin films deposited on SRO electrode and annealed at 500°C in static air for 2 h.

Dielectric Characterization

Figure 3 displayed frequency-dependent dielectric behavior of BLFO film annealed at 500 °C. The dielectric measurements were carried out at room temperature as a function of frequency in the range of 10 kHz - 1 MHz. It is easy to see that the film possess small dielectric dispersion at low frequency since at which the dielectric constants decrease slightly with the frequency. It means that the films has good interface between the BLFO film and SRO bottom electrode. As shown, the dielectric constant shows very little dispersion with frequency indicating that our films possess low defect concentrations at the interface film-substrate. The low dispersion of the dielectric constant and the absence of any relaxation peak in tan δ indicate that both, interfacial polarization of the Maxwell Wagner type and a polarization produced by the electrode barrier can be neglected in the film[19]. The dielectric constant and dissipation factor, at 100 kHz, were found to be 122 and 0.05, respectively. The BLFO film had higher relative dielectric permittivity when compared with those previously reported in ceramics or films[20-25]. The observed improvement of dielectric permittivity may be associated with less structural disorder and less two-dimensional stress in the plane of the film. The tan δ values of the BLFO films keeps unchanged with increasing frequency suggesting that the films exhibited low defect concentration.

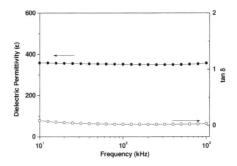

Figure 3. Dielectric permittivity and dielectric loss spectra of BLFO thin films deposited by the polymeric precursor method and annealed at 500° C for 2 hours as function of frequency.

Ferroelectric Characterization

Ferroelectric properties were characterized using both polarization hysteresis as well as pulsed polarization measurements. Figure 4(a) shows a set of hysteresis loops measured at a frequency of 60 Hz. A remnant polarization of Pr = 25 μC/cm^2 is observed, which is smaller than that of films grown on single crystal STO substrates (~55 μC/cm^2 on [100] STO and ~95 μC/cm^2 on [111] STO)[26]. This can be understood as a consequence of the smaller c/a ratio of BLFO on SRO. To more closely investigate the ferroelectric properties, PUND measurements [$\Delta P=P^*$ (switched polarization) - P^ (nonswitched polarization)] were also performed and the result is shown in Figure 4(b). The obtained switched polarization ΔP is 38-42 μC/cm^2 which is consistent with the 2P$_r$ value from the P-E hysteresis loop. The switched polarization values of ~40 μC/cm^2 were observed, which began to saturate at 600-650 kV/cm. Pulse polarization measurements that are less likely convoluted by leakage and nonlinear dielectric effects, confirmed this result. Note also the weak pulse width dependence on Figure 4c once saturation is reached, yet another indication of robust intrinsic ferroelectricity. This demonstrates that the measured polarization switching is an intrinsic property of BLFO thin films, and is not dominated by leakage, which was a critical obstacle in determining the ferroelectric property of bulk BFO[27]. For use in memory applications, the coercive field (which is currently, 2–3 V for 200 nm film) has to be lowered to about 0.7–1 V. Our preliminary experiments using La substitution suggest a similar prospect in the BFO system. The stability of the polar state is confirmed by retention measurements as shown in Figure 4(d). The overall retention time dependence of polarization retention for the BLFO film is quite good. After a retention time of 1×10^4 s, the polarization loss was only about 8.0 % of the value measured at t = 1.0 s for a 9 V applied voltage. For the infant period (within 10 s), depolarization fields could be the main contribution to polarization loss. Similar retention loss behavior has been reported for Bi$_{3.25}$La$_{0.75}$Ti$_3$O$_{12}$ deposited on Pt/TiO$_2$/SiO$_2$/Si[26]. Note that greater than 40% retention loss is often observed for various ferroelectric thin films after 10^4 s even at room temperature[28]. Such behavior has been attributed to a depolarization field which can exist due to the incomplete compensation between the polarization charge and the free charge in the electrodes.

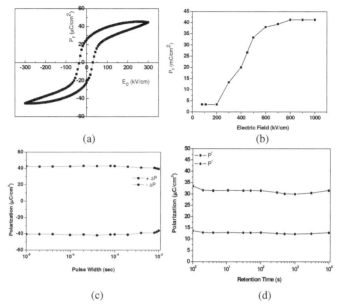

(a)

(b)

(c)

(d)

Figure 4. (a) Remanent polarization -electric field hysteresis loops, (b) Pulsed polarization (ΔP) as a function of an applied voltage and (c) Pulse width dependence of switched polarization (ΔP) in the range from 1 μs to 1 ms, (d) Retention characteristics for BLFO thin film deposited on SRO electrode and annealed at 500°C in static air for 2 h.

Magnetic Characterization

The magnetolectric coefficient versus dc bias magnetic field in the longitudinal and transversal directions reveals hysteretic behavior, as observed in the magnetic field cycles shown in Figures 5a and 5b. The maximum magnetoelectric coefficient of 12 V/cmOe in the longitudinal direction is much larger than that previously reported for thin films as high as 3 V/cmOe in the same direction at zero fields[29]. This is a consequence of the antiferromagnetic axis of BLFO which rotates through the crystal with an incommensurate long-wavelength period of ~ 620 Å[30,31]. Early reports showed that the spiral spin structure leads to a cancellation of any macroscopic magnetization and would inhibit the observation of the linear magnetoelectric effect[32]. Significant magnetization (~ 0.5μB/unitcell) and a strong magnetoelectric coupling have been observed in epitaxial thin films, suggesting that the spiral spin structure could be suppressed [33-35].

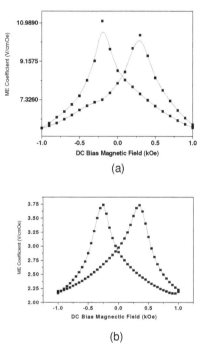

Figure 5. The magnetoelectric coefficient dependence on dc bias magnetic field for BLFO thin films deposited by the polymeric precursor method and annealed at 500° C for 2 hours at a 7 kHz ac magnetic field at room temperature. (a) Longitudinal and (b) Transversal.

Magnetization (*M*) versus field (*H*) loops were recorded at 300 K (Figure 6). The magnetization for the film was observed with a magnetic field of 2.5 emu/g. A weak ferromagnetic response was noted, although enhanced magnetization was observed as compared to bulk specimens. Gehring[36] and Goodenough et al[37] suggested that statistical distribution of Fe^{3+} and Ti^{4+} ions in the octahedra or creation of lattice defects might lead to bulk magnetization and weak ferromagnetism. The appearance of weak ferromagnetism in this compound may be attributed to either the canting of the antiferromagnetically ordered spins by a structural distortion[38] or the breakdown of the balance between the antiparallel sublattice magnetization of Fe^{3+} due to metal ion substitution with a different valence[39,40]. As can be seen, the magnetization of our film linearly increases with the applied magnetic field. This behavior is characteristic of antiferromagnets and can be changed by introducing a small fraction of rare-earth additives. That will change the magnetic structure of pure bismuth ferrite. A change in the magnetic properties in this case is explained not only by different magnetic moments and ion radii of lanthanum (the La atom does not possess any intrinsic magnetic moment), but also by the anisotropy of the magnetic moments of lanthanum ions. Further studies are required to understand the magnetic behaviour of this compound. The low coercive magnetic fields of BLFO film is indicative of its magnetically soft nature and suitablility for device applications.

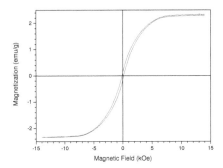

Figure 6. Field dependencies of the magnetization obtained for BLFO films deposited by the soft chemical method and annealed at 500° C for 2 hours.

As can be seen in Figure 7, when films are magnetically poled, initially there is an enhancement in saturation polarization, as expected. However, as the magnetic field increases, the polarization value drops. This may be attributed to disturbance created in grain alignment due the magnetic field. The random orientation of grains can lead to cancellation of polarization values. In that case the reduction in overall P$_s$ value is expected.

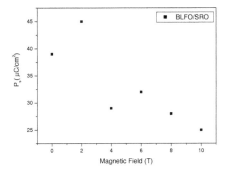

Figure 7. Effect of magnetic poling on saturation polarization (Ps) of BLFO films deposited by the soft chemical method and annealed at 500° C for 2 hours.

Piezoresponse Analysis

The domain structures observed in the film by piezoelectric force microscopy (*PFM*) was illustrated in Figure 8. The out-of-plane (*OP*) and in-plane (*IP*) piezoresponse images of the as-grown films after applying a bias of -12V, on an area of 2 μm x 2 μm, and then an opposite bias of + 12V in the central 1 μm x 1 μm area were employed. For comparison the topography of the film was also analysed (Figure 8a). To obtain the domain images of the films, a high voltage that exceeds the coercive field was applied during scanning. The contrast in these images is associated with the direction of the polarization[41]. The *PFM* image indicates that the perpendicular component of polarization can be switched between two stable states: bright and dark contrast inside and outside of the square region. Higher *PFM* magnification images showed that the regions without piezoresponse exhibit a strong contrast in the *PFM* images. The white regions in the out-of-plane PFM images correspond to domains with the polarization vector oriented toward the bottom electrode hereafter referred to as down polarization (Figure 8b) while the dark regions correspond to domains oriented upward referred to as up polarization. Grains which exhibit no contrast change is associated with zero out-of-plane polarization. A similar situation was observed when a positive bias was applied to the film. We noticed that some of the grains exhibit a white contrast associated to a component of the polarization pointing toward the bottom electrode. On the other hand, in the in-plane *PFM* images (Figure 8c) the contrast changes were associated with changes of the in-plane polarization components. In this case, the white contrast indicates polarization e.g. in the positive direction of the y-axis while dark contrast are given by in-plane polarization components pointing to the negative part of the y-axis. The ferroeletric domains in the BLFO film consist of a multiple domain state in a mixture of 71° and 180° domains which grow large into blocks. The domains grow in multiple states is a consequence of films thickness being close to 300 nm. In this way, lanthanum acts reducing the strain energy and the pinning effect of charged defects. The main differences in the out-of-plane (*OP*) and in-plane (*IP*) piezoresponse images may be understood as follows: First, the piezoelectric tensor for the rhombohedral symmetry is complex, resulting in an effective piezoelectric coefficient that is not proportional to the component of polarization along the detection direction, as explained in Ref.[42]. In this scenario, the *IP* response may not change its sign upon polarization switching, while the *OP* response does. Second, the 180° switching process may take place via two non-180° (i.e., 71°and/or 109°, [43]) switching steps, which also implies switching of only one component of the electrical polarization.

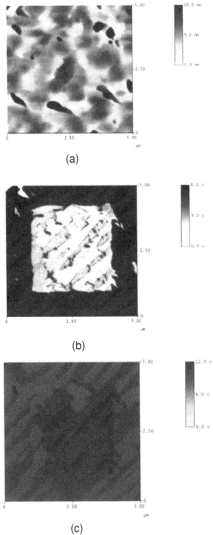

(a)

(b)

(c)

Figure 8. Topography (TP), Out-of-plane (OP) and In-plane (IP) PFM images of BLFO thin films deposited by the polymeric precursor method and annealed at 500° C for 2 hours: (a) TP, (b) (OP) and (c) (IP).

The d$_{33}$(V) hysteresis loop is shown in Figure 9. The maximum d$_{33}$ value, ~80 pm/V is better than the reported value for a BFO deposited on Si[44]. The enhancement of polarization could be caused by the (111) orientation of the ferroelectric film deposited on SRO oxide which reduces the initial nucleation rate when crystallizing the film. The presented value reported for our BLFO film suggests that this material can be considered as a viable alternative for lead-free piezo-ferroelectric devices.

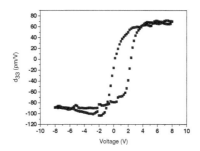

Figure 9. Piezoelectric coefficient loop, $d_{33,}$ of BLFO thin film deposited by the polymeric precursor method on SRO electrode and annealed at 500°C in static air for 2 h.

CONCLUSIONS

Magnetoelectric effect in BLFO films grown on the SRO buffered Pt coated silicion substrates by the soft chemical method was noted. The maximum magnetoelectric coefficient was close to 12 V/cmOe in the longitudinal direction. The high dielectric permittivity of the BLFO films was mainly due to the less structural disorder and less two-dimensional stress in the plane of the film. Ferroelectric domain characteristics by PFM were also investigated to study the dependence of electrical properties on grain orientations. A simple analysis based on three types of contrast, namely, bright, dark and gray contrast was used to determine domain configuration. It was found that 71° and 180° domain switching's, and pinned domain formation can occur in BLFO thin films. In-plane magnetization-field curves revealed magnetization of the BLFO films. Piezoelectric force microscopy images reveal that *in-plane* response may not change its sign upon polarization switching, while the *out-of-plane* response does. Lanthanum substitution was found to effectively induce spontaneous magnetization in antiferromagnetic BiFeO₃ exhibiting good piezoelectric properties. The presence of magnetoelectric coupling at room temperature is noteworthy since it could give additional degrees of freedom in device fabrication.

ACKNOWLEDGEMENT

The financial support of this research project by the Brazilian research funding agencies CNPq and FAPESP is gratefully acknowledged.

REFERENCES

[1] G. A. Smolenskii and I. E. Chupis, Sov. Phys. Usp. **25**, 475-479 (1982).
[2] W. Eerenstein, F. D. Morrison, J. Dho, M. G. Blamire, and J. F. Scott, Science. **307**, 1203-1207 (2005).
[3] A. Vasudevarao et al., Phys. Rev. Lett. **97**, 257602 -257605 (2006).
[4] K. Kato and S. Iida, J. Phys. Soc. Jpn. **51,** 1335-1338 (1982).
[5] Y. Yamasaki et al., Phys. Rev. Lett. **96,** 207204-207206 (2006).

[6] M. Gajek et al., Nature Mater. **6,** 296-299 (2007) .

[7] W. J. Gallagher and S. S. P. Parkin, IBM J. Res. Develop. **50,** 5 (2006).

[8] M. Dawber, K. M. Rabe, and J. F. Scott, Rev. Mod. Phys. **77,** 083-087 (2005).

[9] P. Royen and K. Swars, Angew. Chem. **24** 779-781 (1957).

[10] S. V. Kiselev, R. P. Ozerov, and G. S. Zhdanov, Sov. Phys. Dokl. **7,** 742-745 (1963).

[11] I. Sosnowksa, T. Peterlin-Neumaier, and E. Steichele, J. Phys. C. **15**, 4835-4839 (1982).

[12] J. Lu, D.-A. Pan, and L. Qiao, e-print ar Xiv:0704/0704.2990.

[13] A. Z. Simões, A. H. M. Gonzalez, E.C. Aguiar, C.S. Riccardi, E. Longo, J.A.Varela, Appl. Phys. Lett. **93,** 142902-142905 (2009).

[14] J. Wang, J. B. Neaton, H. Zheng, V. Nagarajan, S. B. Ogale, B. Liu, D. Viehland, V. Vaithyanathan, D. G. Schlom, U. V. Waghmare, N. A. Spaldin, K. M. Rabe, M. Wuttig, and R. Ramesh, Science. **299,** 1719-1722 (2003).

[15] A. Z. Simões, L. S. Cavalcante, C. S. Riccardi, J. A. Varela, and E. Longo, Curr. Appl. Phys. **9,** 520-523 (2009)

[16] A.Z.Simões, M.A. Ramirez, C.S. Riccardi, A. Ries, E. Longo, J.A. Varela, Materials Chemistry and Physics. **92**, 373-378 (2005).

[17] Th. Schedel-Niedrig, W. Weiss, and R. Schlögl, Phys. Rev. B 52, 17449-17460 (1995).

[18] J. Li, J. Wang, M. Wuttig, R. Ramess, N. Wang, B. Ruette, A. P. Pyatakov, A. K. Zvezdin, and D. Viehland, Appl. Phys. Lett. **84,** 5261-5263 (2004).

[19] V. R. Palkar, Darshan C. Kundaliya, S. K. Malik, and S. Bhattacharya, Phys. Rev. B **69**, 212102-212104 (2004).

[20] R. Ramesh and N. A. Spaldin, Nat. Mater. **6,** 21-24 (2007).

[21] I. Sosnowska, T. Peterlin-Neumaier, and E. Streichele, J. Phys. C. **15**, 4835-4838 (1982).

[22] Y. F. Popov, A. K. Zvezdin, G. P. Vorbev, A. M. Kadomtseva, V. A. Murashev, and D. N. Racov, JETP Lett. **57,** 69-71 (1993).

[23] F. Bai, J. Wang, M. Wuttig, J. F. Li, N. Wang, A. P. Pyatakov, A. K. Zvezdin, L. E. Cross, and D. Viehland, Appl. Phys. Lett. **86,** 032511-032513 (2005).

[24] T. Kojima, T. Sakai, T. Watanabe, H. Funakubo, Appl. Phys. Lett. 80 (2002) 2746-2748.

[25] G. L. Yuan, S. W. Or, Y. P. Wang, Z. G. Liu, J. M. Liu, Solid State Commun. 138, 76 **(2006).**

[26] J. F. Li, J. Wang, M. Wuttig, R. Ramesh, N. Wang, B. Ruette, A. P. Pyatakov, A. K. Zvezdin, and D. Viehland, Appl. Phys. Lett. **84,** 5261-5263 (2004).

[27] J. R. Teague, R. Gerson, and W. J. James, Solid State Commun. **8**, 1073-1074 (1970).

[28] B. S. Kang, J.-G. Yoon, T. K. Song, S. Seo, Y. W. So, and T. W. Noh, Jpn. J. Appl. Phys. **41,** 5281-5283 (2002).

[29] J. W. Hong, W. Jo, D. C. Kim, S. M. Cho, H. J. Nam, H. M. Lee, and J.U. Bu, Appl. Phys. Lett. **75,** 3183-3185 (1999).

[30] J. Li, J. Wang, M. Wuttig, R. Ramess, N. Wang, B. Ruette, A. P. Pyatakov, A. K. Zvezdin, and D. Viehland, Appl. Phys. Lett. **84,** 5261-5263 (2004).

[31] R. Ramesh and N. A. Spaldin, Nat. Mater. **6,** 21-25 (2007).

[32] I. Sosnowska, T. Peterlin-Neumaier, and E. Streichele, J. Phys. C. **15,** 4835-4839 (1982).

[33] Y. F. Popov, A. K. Zvezdin, G. P. Vorbev, A. M. Kadomtseva, V. A. Murashev, and D. N. Racov, JETP Lett. **57,** 69 (1993).

[34] F. Bai, J. Wang, M. Wuttig, J. F. Li, N. Wang, A. P. Pyatakov, A. K. Zvezdin, L. E. Cross, and D. Viehland, Appl. Phys. Lett. **86,** 032511-032513 (2005).

[35] M. M. Kumar, A. Srinivas and S. V. Suryanarayana J. Appl. Phys. **87,** 855-862 (2000).

[36] G. A. Gehring, Ferroelectrics. **161,** 275-279 (1994).

[37] J. B. Goodenough and J. M. Lango Landolt–Bornstein Numerical Data and Functional Relationships in Science and Technology vol III/4a (New York: Springer, 1978)

[38] M. M. Kumar, S. Srinath, G. S. Kumar and S. V. Suryanarayana, J. Magn. Magn. Mater. **188,** 203-212 (1998).

[39] T. Kanai, S. I. Ohkoshi, A. Nakajima, T. Watanabe and K. Hashimoto Adv. Mater. **13**, 487-490 (2001).

[40] J. Wang, A. Scholl, H. Zheng, S. B. Ogale, L. Viehland, D. G. Schlom, N. A. Spaldin, K. M. Rabe, M.Wuttig, L. Mohaddes, J. Neaton, U. Waghmare, T. Zhao and R. Ramesh, Science **307**, 1203-1206 (2005).

[41] K.Ueda, H. Tabata and T. Kawai Appl. Phys. Lett. **75**, 555-557 (1999).

[42] T. Kanai, S. I. Ohkoshi and K. Hashimoto J. Phys. Chem. Solids **64**, 391-397 (2003).

[43] T. Kojima, T. Sakai, T. Watanabe, H. Funakubo, Appl. Phys. Lett. **80**, 2746-2748 (2002).

[44] J. Wang, J. B. Neaton, H. Zheng, D. G. Schlom, U. V. Waghmare, N. A. Spaldin, K. M. Rabe, M. Wuttig, R. Ramesh, Science. 299, 1719-1722 (2003).

PROPERTIES OF $Pb(Zr,Ti)O_3/CoFe_2O_4/Pb(Zr,Ti)O_3$ LAYERED THIN FILMS PREPARED VIA CHEMICAL SOLUTION DEPOSITION

Yoshikatsu Kawabata, Makoto Moriya, Wataru Sakamoto and Toshinobu Yogo
Division of Nanomaterials Science, EcoTopia Science Institute, Nagoya University
Furo-cho, Chikusa-ku, Nagoya 464-8603, Japan

ABSTRACT

Multiferroic $Pb(Zr,Ti)O_3/CoFe_2O_4/Pb(Zr,Ti)O_3$ layered thin films were prepared on $Pt/TiO_x/SiO_2/Si$ substrates via chemical solution deposition using metal-organic precursor solutions. Layered composite films composed of perovskite-structured $Pb(Zr_{0.52}Ti_{0.48})O_3$ and spinel-structured $CoFe_2O_4$ phases without any secondary phases were successfully synthesized by controlling the processing conditions. The layered structure of the $Pb(Zr_{0.52}Ti_{0.48})O_3/CoFe_2O_4/Pb(Zr_{0.52}Ti_{0.48})O_3$ was also confirmed via scanning electron microscope observations. These layered films simultaneously exhibited sufficiently high insulating properties and ferroelectric polarization-electric field and ferromagnetic magnetization-magnetic field hysteresis loops at room temperature. Furthermore, on the basis of the magnetoelectric interaction between the layers, changes in their ferroelectric properties were observed when a magnetic field (approximately 0.5 T) was applied.

INTRODUCTION

Multiferroic magnetoelectric (ME) materials that simultaneously exhibit both ferroelectricity and ferromagnetism have recently stimulated an increasing number of research activities because of interest in their properties.[1,2] ME composite materials that contain a combination of ferroelectric and ferromagnetic phases[3-8] are more feasible for practical applications than $BiFeO_3$-based single-phase multiferroic materials because of their greater ME coupling at room temperature[9-11]. In this case, the selection of materials with excellent electric-field-induced strain and magnetostrictive properties for the ferroelectric and ferromagnetic phases is important for the fabrication of novel ME composites. Furthermore, among several ME composites, layer-structured thin films that consist of ferroelectric and ferromagnetic layers have been receiving significant attention because of the magnitude of their elastic coupling interaction.[2-5] As ferroelectric and ferromagnetic substances for such a composite, $Pb(Zr,Ti)O_3$ and $CoFe_2O_4$, respectively, are considered to be suitable materials because they exhibit superior electrical and magnetic characteristics with higher Curie temperatures compared to other available materials. The $Pb(Zr,Ti)O_3$ ceramics, especially at the morphotropic phase boundary composition of $Pb(Zr_{0.52}Ti_{0.48})O_3$, exhibit excellent dielectric and piezoelectric properties.[12] In addition, at approximately 0.6 T, polycrystalline $CoFe_2O_4$ has been reported to exhibit a large magnetostrictive strain of approximately -130 ppm in the direction parallel to that of the applied magnetic field.[13]

However, to realize the ME effect by the formation of a layered (2-2 type) composite film of these materials, it is also important and indispensable that high-quality samples of such films be fabricated using a simple and easy technique. Chemical solution deposition (CSD) is one of the well-known methods for the fabrication of thin films on substrates. This method is a useful and suitable process for achieving high homogeneity, a lower processing temperature, precise control of the chemical composition, and versatile shapes with low equipment costs.

In this study, we selected $Pb(Zr_{0.52}Ti_{0.48})O_3$ (PZT) and $CoFe_2O_4$ (CFO) as the ferroelectric and ferromagnetic phases, respectively, for the fabrication of layer-structured films via CSD. Tailored metal-organic precursor solutions for the preparation of PZT and CFO layers were alternately spin-coated onto Si-based substrates to fabricate PZT/CFO/PZT layered films. In this case, a triple-layered structure was selected to achieve superior insulating properties compared to a Pb(Zr,Ti)O$_3$/CoFe$_2$O$_4$ bilayered film. These superior insulating properties result from the fact that magnetic CFO layers usually exhibit an insulating resistance much lower than dielectric PZT layers. The effect of processing conditions on the crystallographic phase, the microstructure, and several other properties of the PZT/CFO/PZT composite films were studied to demonstrate the feasability of preparing the ME thin film on Si-based substrates.

EXPERIMENTAL DETAILS

Pb(OCOCH$_3$)$_2$, Zr(O^{i}C$_3$H$_7$)$_4$ and Ti(O^{i}C$_3$H$_7$)$_4$ were selected as the starting materials for the preparation of a PZT precursor solution. 2-Methoxyethanol was used as a solvent; prior to use, it was dried over molecular sieves and distilled. The procedure was conducted under a dry N$_2$ atmosphere because the starting materials are extremely sensitive to moisture. Amounts of Pb(OCOCH$_3$)$_2$, Zr(O^{i}C$_3$H$_7$)$_4$, and Ti(O^{i}C$_3$H$_7$)$_4$ that corresponded to a composition of Pb(Zr$_{0.52}$Ti$_{0.48}$)O$_3$ with a 5 mol% excess of Pb were dissolved in absolute 2-methoxyethanol and refluxed for 18 h to yield a homogeneous precursor solution (0.3 mol/l). In this case, acetylacetone was added to the solution as a stabilizing agent. The molar ratio of acetylacetone to the PZT precursor was set at 4.0.

For the preparation of a CFO precursor solution, Co(O^{i}C$_3$H$_7$)$_2$ and Fe(OC$_2$H$_5$)$_3$ were used as the starting materials. Amounts of Co(O^{i}C$_3$H$_7$)$_2$ and Fe(OC$_2$H$_5$)$_3$ that corresponded to a composition of CoFe$_2$O$_4$ were dissolved in absolute 2-methoxyethanol and refluxed for 18 h under a dry N$_2$ atmosphere. A homogeneous CFO precursor solution (0.05 mol/l) was obtained. Acetylacetone was also added to the solution in the same molar ratio used for the PZT precursor solution (acetylacetone/CFO precursor = 4.0 (molar ratio)).

PZT and CFO precursor films were fabricated by spin coating the PZT and CFO precursor solutions on Pt/TiO$_x$/SiO$_2$/Si substrates. The precursor film was dried at 150°C for 5 min, and then calcined at 400°C for 1 h (heating rate: 5°C/min) in an O$_2$ flow. After five drying and calcining cycles, the film was crystallized at 650–700°C for 30 min in an O$_2$ flow using a conventional tube furnace (10°C/min). To fabricate PZT/CFO/PZT layered films, five coating layers of PZT and CFO were alternately deposited as a single unit. The detailed coating and heat-treatment schemes are shown in Fig. 1. The coating and calcining cycles were repeated 15 times (10 times for PZT and five times for CFO) with three heat treatments for the crystallization of each PZT and CFO layer, resulting in a layered PZT/CFO/PZT film thickness of approximately 600 nm.

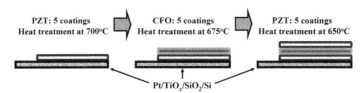

Figure 1. Process flow for the fabrication of the PZT/CFO/PZT layered thin film via CSD: coating and heat-treatment schemes.

The crystallographic phases of the thin films were examined via X-ray diffraction (XRD) analysis using monochromated Cu-Kα radiation. The microstructure of the thin films was characterized using a field-emission scanning electron microscope (FE-SEM). The electrical properties of the PZT/CFO/PZT thin films were measured using Pt top electrodes with a diameter of 0.2 mm deposited via DC sputtering onto the surface of the upper PZT layer followed by annealing at 400°C for 1 h. The Pt layer of the substrate was used as the bottom electrode. The ferroelectric properties of the films were evaluated using a ferroelectric test system at room temperature. The leakage current properties of the PZT/CFO/PZT layered thin films were measured using an electrometer/high-resistance meter at room temperature. The magnetic properties of the crystalline PZT/CFO/PZT layered film were characterized using a vibrating sample magnetometer (VSM) at room temperature.

RESULTS AND DISCUSSION

Figure 2(a) shows the XRD pattern of a PZT/CFO/PZT layered thin film fabricated on a Pt/TiO$_x$/SiO$_2$/Si substrate. The diffraction profile of the layered thin film contained the diffraction peaks corresponding to the perovskite Pb(Zr,Ti)O₃ with a random orientation. In addition, diffraction peaks that correspond to the polycrystalline CoFe₂O₄ were also observed. The XRD pattern of the crystallized PZT/CFO/PZT composite thin film revealed that all of the diffraction peaks belonged to the perovskite-structured Pb(Zr,Ti)O₃ and to the spinel-structured CoFe₂O₄, except for those attributable to the substrate. No peak that corresponds to the additional or intermediate phase was observed. On the basis of the results in Fig. 2(a), composite thin films with PZT and CFO layers were successfully synthesized by controlling the coating and heating conditions of each PZT and CFO precursor layer, as shown in Fig. 1.

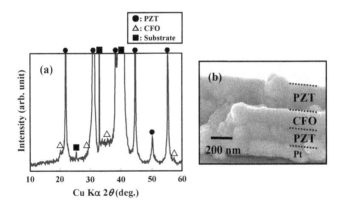

Figure 2. (a) XRD pattern and (b) Cross-sectional SEM image of a PZT/CFO/PZT thin film fabricated via CSD, as shown in Fig. 1.

Figure 2(b) shows a cross-sectional SEM image of a PZT/CFO/PZT layered thin film prepared on a Pt/TiO$_x$/SiO$_2$/Si substrate. This film exhibited a smooth surface morphology and had a uniform thickness. The SEM micrograph of the composite thin film revealed clear interfaces for not only the substrate but also the interfaces between $Pb(Zr,Ti)O_3$ and $CoFe_2O_4$. The formation of the PZT/CFO/PZT layered structure was confirmed, as shown in Fig. 2(b). The thickness of the PZT layers was approximately 450 nm (upper and bottom PZT layers) and that of the intermediate CFO layer was approximately 150 nm.

The synthesized PZT/CFO/PZT-structured films exhibited leakage current densities less than 10^{-6} A/cm^2 even at high applied fields (>200 kV/cm) at room temperature. For reference, a CFO (five coatings) single-layer film without upper and bottom PZT layers, PZT (five coatings)/CFO (five coatings) and PZT (10 coatings) films without upper PZT and intermediate CFO layers, respectively, were also fabricated, and their leakage current properties were evaluated. The PZT/CFO bilayered film exhibited leakage current densities at high applied fields (>200 kV/cm) that were 1–2 orders of magnitude greater than those of the PZT/CFO/PZT films. This result was because of the leakage current densities of CFO single-layer films being approximately 5–6 orders of magnitude greater than those of PZT single-layer films over a wide range of applied fields. In addition, the leakage current properties of the PZT/CFO/PZT thin films were consistent with those of the PZT single-layer film. Thus, the insulating resistance of the PZT/CFO/PZT films was sufficiently high to allow their ferroelectric properties to be evaluated.

Figure 3 shows the polarization (P)–electric field (E) hysteresis loops of PZT single-layer and PZT/CFO/PZT layered thin films measured at room temperature. In this case, the applied electric field of the PZT/CFO/PZT thin film was corrected according to the thickness of the PZT layers in PZT/CFO/PZT because the electrical resistivity of the CFO layer is considerably lower than that of the PZT layer; with the correction, the electric fields were effectively applied on the PZT layers. As shown in Fig. 3, a PZT single-layer film exhibited a well-shaped ferroelectric hysteresis loop with a small coercive field of less than 100 kV/cm. The values of the remnant polarization (P_r) and the coercive field (E_c) of the PZT and PZT/CFO/PZT thin films were approximately 32 and 30 μC/cm^2 and 90 and 240 kV/cm^2, respectively. The PZT/CFO/PZT layered thin film also exhibited a well-shaped ferroelectric hysteresis loop and a remanent polarization of approximately 30 μC/cm^2. Although the PZT/CFO/PZT film showed a P_r value comparable to that of the PZT single-layer film, relatively larger coercive fields were observed. This was because of the existence of a CFO intermediate layer (i.e., a clamping effect and a reduction of the effective applied field), which affected the domain switching of the ferroelectric PZT layer under an applied electric field.

The magnetic properties of the PZT/CFO/PZT films were also evaluated using a VSM. Figure 4 shows the magnetization (M)–magnetic field (H) hysteresis loop of a PZT/CFO/PZT thin film on a $Pt/TiO_x/SiO_2/Si$ substrate. This measurement was performed at room temperature. The magnetization was normalized only to the volume of the CFO layer. Furthermore, the data was also corrected for the diamagnetism component from the substrate and sample holder. The PZT/CFO/PZT sample exhibited typical ferromagnetic characteristics. The typical M-H hysteresis loop of ferromagnetism shown in Fig. 4 had a magnetization at an applied field of 16 kOe, a remanent magnetization, and a coercive force of approximately 200 emu/cc, 95 emu/cc, and 1.2 kOe, respectively. A saturation magnetization value comparable to the reported for a sol-gel-derived $CoFe_2O_4$ thin film[5] was observed for the film. However, the obtained data revealed properties inferior to those of bulk $CoFe_2O_4$. Several factors, such as the size and surface effects associated with the small grain size (approximately 100 nm, which was confirmed by atomic force microscopy) as well as the defect structure of the CFO layer, also affected the magnetic properties (i.e., the magnetization behavior and coercive force). Furthermore, the stress derived from the substrate that results from the mismatch of the thermal expansion coefficient between the film and substrate should be considered. The effects of these factors on the magnetic properties of the layered composite film are still not clear and are currently under investigation.

On the basis of the results shown in Fig. 3, the PZT/CFO/PZT thin film was determined to be multiferroic at room temperature. In the case of single-phase multiferroic materials, single-crystalline $BiFeO_3$ has been reported to exhibit a near-linear M-H relationship.[10] Polycrystalline $BiFeO_3$ samples also showed very small magnetizations and their M-H hysteresis loops did not show spontaneous magnetization.[14] For the current PZT/CFO/PZT layered thin-film sample, magnetization significantly greater than that for the previously reported $BiFeO_3$-based thin films was observed.[10,11,15,16] The greater magnetization of the PZT/CFO/PZT thin films is attributed to the presence of the ferromagnetic CFO layer, even though its volume fraction in the film is relatively small.

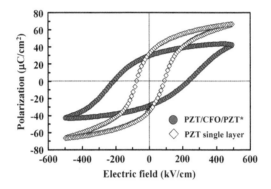

Figure 3. P−E hysteresis loops of PZT and PZT/CFO/PZT layered thin films:
  ◇ PZT single-layer thin film crystallized at 700°C (10 coatings)
  ● PZT/CFO/PZT layered thin film*
  [Frequency of measurement: 1000 Hz]
  *The applied electric field was corrected according to the thickness of the PZT layers.

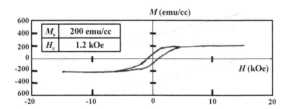

Figure 4. M−H hysteresis curve of a PZT/CFO/PZT layered thin film*
  *The magnetization was normalized only to the volume of the CFO layer.

The ME interaction of the PZT/CFO/PZT films was also evaluated via ferroelectric measurements performed under a magnetic field. Figure 5 shows the ferroelectric P-E hysteresis loops of a PZT/CFO/PZT thin film on a Pt/TiO$_x$/SiO$_2$/Si substrate in both the presence and absence of an applied magnetic field. For this measurement, the layered film was placed on an Nd$_2$Fe$_{14}$B-based magnet with a surface magnetic flux density of 0.5 T. The polarization value slightly decreased upon application of the magnetic field. Given the magnetostrictive properties of CoFe$_2$O$_4$, the CFO layer might exhibit a positive strain in the plane direction under the magnetic field because polycrystalline CoFe$_2$O$_4$ has been reported to show a magnetostrictive strain of approximately +110 ppm perpendicular to the direction of the applied magnetic field at approximately 0.5 T.[13] In this case, tensile stress (magnetostrictive strain-induced stress) was applied to the PZT layers. The polarization properties of the PZT layers should be diminished under the magnetic field. For comparison, a similar tendency in a past report[17] is provided in Fig. 5. However, the reason for the behavior shown in Fig. 5 is still not clear. Furthermore, the measurement of the induced ME voltage properties is necessary for the current composite films. The detailed mechanism and the optimization of PZT and CFO thicknesses that provide the improved ME response should be studied in future work.

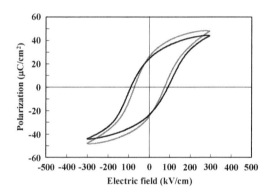

Figure 5. P–E hysteresis loops for a PZT/CFO/PZT layered thin film* in the presence and in the absence of an applied magnetic field (0.5 T)
[Frequency of measurement: 100 Hz].
(The applied magnetic field direction is perpendicular to the film plane.)
*The applied electric field was corrected according to the thickness of the PZT layers.

CONCLUSIONS

Magnetoelectric (ME) Pb(Zr$_{0.52}$Ti$_{0.48}$)O$_3$/CoFe$_2$O$_4$/Pb(Zr$_{0.52}$Ti$_{0.48}$)O$_3$ layered thin films were successfully synthesized via chemical solution deposition using metal-organic precursor solutions. In this process, both Pb(Zr$_{0.52}$Ti$_{0.48}$)O$_3$ and CoFe$_2$O$_4$ precursor solutions were alternately spin-coated onto Pt/TiO$_x$/SiO$_2$/Si substrates to form layered composite films and the heating conditions for each Pb(Zr$_{0.52}$Ti$_{0.48}$)O$_3$ and CoFe$_2$O$_4$ precursor layer were appropriately controlled. XRD patterns of the crystallized Pb(Zr$_{0.52}$Ti$_{0.48}$)O$_3$/CoFe$_2$O$_4$/Pb(Zr$_{0.52}$Ti$_{0.48}$)O$_3$

composite thin films revealed that all of the diffraction peaks corresponded to perovskite-structured $Pb(Zr_{0.52}Ti_{0.48})O_3$ and the spinel-structured $CoFe_2O_4$. No peaks that corresponded to secondary phases were observed. From the cross-sectional SEM images of the composite thin films, clear interfaces between the $Pb(Zr_{0.52}Ti_{0.48})O_3$ and $CoFe_2O_4$ layers were observed. These layered films exhibited sufficient insulating properties, and ferroelectric and ferromagnetic hysteresis loops were also simultaneously observed at room temperature. Furthermore, from the ferroelectric measurements under an applied magnetic field (approximately 0.5 T), changes in the polarization behavior caused by their magnetoelectric interaction were observed for the layered films.

REFERENCES

[1] W. Eerenstein, N. D. Mathur, and J. F. Scott, Multiferroic and Magnetoelectric Materials, *Nature*, **442**, 759–765 (2006).

[2] L. W. Martin, Y.-H. Chu, and R. Ramesh, Advances in the Growth and Characterization of Magnetic, Ferroelectric, and Multiferroic Oxide Thin Films, *Mater. Sci. Eng. R*, **68**, 89–133 (2010).

[3] H.-C. He, J.-P. Zhou, J. Wang, and C.-W. Nan, Multiferroic $Pb(Zr_{0.52}Ti_{0.48})O_3$–$Co_{0.9}Zn_{0.1}Fe_2O_4$ Bilayer Thin Films via a Solution Processing, *Appl. Phys. Lett.*, **89**, 052904 (2006).

[4] C.-W. Nan, M. I. Bichurin, S. Dong, D. Viehland, and G. Srinivasan, Multiferroic Magnetoelectric Composites: Historical Perspective, Status, and Future Directions, *J. Appl. Phys.*, **103**, 031101 (2008).

[5] H.-C. He, J. Ma, J. Wang, and C.-W. Nan, Orientation-Dependent Multiferroic Properties in $Pb(Zr_{0.52}Ti_{0.48})O_3$–$CoFe_2O_4$ Nanocomposite Thin Films Derived by a Sol-Gel Processing, *J. Appl. Phys.*, **103**, 034103 (2008).

[6] C. G. Zhong, Q. Jiang, J. H. Fang, and X. F. Jiang, Thickness and Magnetic Field Dependence of Ferroelectric Properties in Multiferroic $BaTiO_3$–$CoFe_2O_4$ Nanocomposite Films, *J. Appl. Phys.*, **105**, 044901 (2009).

[7] W. Bai, X. Meng, T. Lin, X. Zhu, J. Ma, W. Liu, J. Sun, and J. Chu, Magnetic Field Induced Ferroelectric and Dielectric Properties in $Pb(Zr_{0.5}Ti_{0.5})O_3$ Films Containing Fe_3O_4 Nanoparticles, *Thin Solid Films*, **518**, 3721–3724 (2010).

[8] L.W. Martin and R. Ramesh, Multiferroic and Magnetoelectric Heterostructures, *Acta Mater.*, **60**, 2449–2470 (2012).

[9] J. Wang, J. B. Neaton, H. Zheng, V. Nagarajan, S. B. Ogale, B. Liu, D. Viehland, V. Vaithyanathan, D. G. Schlom, U. V. Waghmare, N. A. Spaldin, K. M. Rabe, M. Wuttig, and R. Ramesh, Epitaxial $BiFeO_3$ Multiferroic Thin Film Heterostructures, *Science*, **299**, 1719-1722 (2003).

[10] F. Bai, J. Wang, M. Wittig, J.-F. Li, N. Wang, A. P. Pyatakov, A. K. Zvezdin, L. E. Cross, and D. Viehland, Destruction of Spin Cycloid in $(111)_c$-Oriented $BiFeO_3$ Thin Films by Epitaxial Constraint: Enhanced Polarization and Release of Latent Magnetization, *Appl. Phys. Lett.*, **86**, 032511 (2005).

[11] X. Zhang, Y. Sui, X. Wang, J. Mao, R. Zhu, Y. Wang, Z. Wang, Y. Liu, and W. Liu, Multiferroic and Magnetoelectric Properties of Single-Phase $Bi_{0.85}La_{0.1}Ho_{0.05}FeO_3$ Ceramics, *J. Alloys Compd.*, **509**, 5908–5912 (2011).

[12] B. Jaffe, W. R. Cook, and H. Jaffe, Piezoelectric Ceramics, *Academic Press Inc.*, NewYork, 1971.

[13]S. D. Bhame and P. A. Joy, Tuning of the Magnetostrictive Properties of CoFe$_2$O$_4$ by Mn Substitution for Co, *J. Appl. Phys.*, **100**, 113911 (2006).

[14]A. K. Pradhan, K. Zhang, D. Hunter, J. B. Dadson, G. B. Loutts, P. Bhattacharya, R. Katiyar, J. Zhang, D. J. Sellmyer, U. N. Roy, Y. Cui, and A. Burger, Magnetic and Electrical Properties of Single-Phase Multiferroic BiFeO$_3$, *J. Appl. Phys.*, **97**, 093903 (2005).

[15]W. Sakamoto, A. Iwata, M. Moriya, and T. Yogo, Electrical and Magnetic Properties of Mn-Doped 0.7BiFeO$_3$–0.3PbTiO$_3$ Thin Films Prepared Under Various Heating Atmospheres, *Mater. Chem. Phys.*, **116**, 536-541 (2009).

[16]A. Hieno, W. Sakamoto, M. Moriya, and T. Yogo, Synthesis of BiFeO$_3$–Bi$_{0.5}$Na$_{0.5}$TiO$_3$ Thin Films by Chemical Solution Deposition and Their Properties, *Jpn. J. Appl. Phys.*, **50**, 09NB04 (2011).

[17]M. Tsukada, H. Yamawaki, and M. Kondo, Crystal Structure and Polarization Phenomena of Epitaxially Grown Pb(Zr,Ti)O$_3$ Thin-Film Capacitors, *Appl. Phys. Lett.*, **83**, 4393–4395(2003).

INTELLIGENT PROCESSES ENABLE NEW PRODUCTS IN THE FIELD OF NON-OXIDE CERAMICS

Jens Eichler
ESK Ceramics GmbH & Co. KG
Kempten, Germany

ABSTRACT

Heat exchangers and flow reactors made from silicon carbide pave the way for the modern synthesis of valuable fine chemicals and pharmaceuticals under controlled and safe conditions. The production of such apparatus includes our developed process for joining SiC into complex monolithic shapes. By integrating the ceramic into a system it makes for resource- and energy-efficient processes in the target industries.

An overview is given of the processing steps involved in the production of silicon-carbide heat exchangers and flow reactors, including powder synthesis, powder processing, forming, sintering and assembly. For each of these processing steps, the current state of development and challenges are addressed from our perspective.

INTRODUCTION

We develop and manufacture pioneering products in the fields of advanced ceramics, ceramic powders and frictional coatings. Our non-oxide materials portfolio includes borides, nitrides and carbides, such as silicon carbide (SiC), titanium diboride (TiB$_2$), and boron nitride (BN). In recent years, we have developed new materials and applications together with our customers. We continually develop innovative product solutions and employ a wide range of process technologies.

EKasic® Silicon Carbide Modular Flow Reactor Systems are a new, versatile and efficient chemical manufacturing system. Conventional reactors reach their limits in corrosive media and high-temperature applications. To allow operation under such conditions, we develop, produce and market continuous flow reactors made of silicon carbide engineering ceramics. EKasic® Silicon Carbide Plate Heat Exchangers combine maximum heat transfer with minimum pressure loss. Our developed diffusion bonding technology enables the production of gas-tight monolithic silicon carbide apparatus with complex three-dimensional channel systems. By integrating the ceramic into a system, it makes for resource- and energy-efficient processes in the target industries.

Below, typical processing steps for non-oxide ceramics are discussed with the examples of flow reactors and heat exchangers. In an outlook, the author raises questions from our perspective concerning future trends in processing and manufacturing technology.

APPLICATION EXAMPLE: HEAT EXCHANGERS AND FLOW REACTORS

The material property fingerprint for the application as heat exchangers and flow reactors reveals the need for a material that combines excellent corrosion resistance against aggressive acids and alkalis with high thermal conductivity and thermal stability, good strength and pressure resistance, and in some cases high abrasion resistance.[1] Solid-state sintered silicon carbide offers this combination of properties, thus enabling a new class of heat exchangers and flow reactors for challenging conditions.

The combination of thermal conductivity (> 110 W/mK), thermal stability (up to 1500°C), good strength (especially compared to glass and polymers) and very good chemical resistance make solid-state sintered silicon carbide the ideal material for producing gas-tight flow reactors to control fast, highly exothermic reactions or endothermic reactions with high enthalpy. The high abrasion resistance of silicon carbide, due to its excellent hardness, is an additional advantage if abrasive particle are to be processed. The monolithic design and high fracture strength of EKasic® silicon carbide allow operating pressure up to 16 bar in (one of the) standard heat exchanger configurations. Special customized heat exchanger models have even been operated up to 60 bar at temperatures exceeding 200°C.

Custom-made silicon carbide plates are joined by our developed process to form a monolithic apparatus that can be combined with a frame and connections to form a flow reactor or heat exchanger. The apparatus is a plug-in system for performing the necessary reactions or processes at our customer's facilities (see Figure 5).

Producing heat exchangers and flow reactors involves silicon carbide powder synthesis, powder processing, forming, sintering and final assembly (Figure 1). Each of these steps is critical to the final quality of the complete system. To further optimize the processing of complex ceramic systems, innovations must be made in each processing step. This section gives an overview of the processes that are currently typically applied and gives an outlook for further improvements that may be necessary.

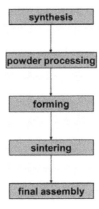

Figure 1. Processing steps for heat exchangers and flow reactors (source ESK)

SYNTHESIS

The synthesis of silicon carbide by carbothermal reaction by the Acheson process is a well-established technology. Theoretically, a stoichiometric mixture of 150 t SiO_2 and 90 t carbon will react at 2500°C for several days to form silicon carbide. Due to the oxidation of carbon and the use of raw material with lower purity, the mixture is adjusted to 100 t SiO_2 and 140 t carbon as recommended in the literature.[2,3] 7.8-8 kWh of electric energy is consumed per kg of the final SiC product. The crystal structure (alpha or beta) of the final product depends on the processing temperature. The color of the SiC crystals depends on the impurities present in the crystal structure (nitrogen – green; aluminum – blue/black).[2,4]

One advantage of silicon carbide synthesis is that the raw materials used are available worldwide and make up a major proportion of the Earth's crust. Therefore, the widespread availability and low cost of the raw materials should make silicon carbide an even more attractive material in the future, as other raw materials become rare or inaccessible.

Today, the total world production of one million tons of SiC per year at a powder price of 1-5 EUR/kg makes SiC attractive for applications that must withstand challenging conditions.[4,5]

But the high energy consumption for the synthesis of silicon carbide could make the widespread use of this material somewhat difficult in the future.

POWDER PROCESSING

Powder processing includes the steps of milling, chemical treatment and fabrication of granules with good processability. Figure 2 shows the typical process route to sintered silicon carbide, as used e.g. for heat exchanger plates or in flow reactors.

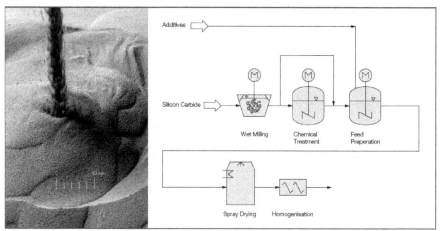

Figure 2. Fabrication of spray granules with good processability (source ESK).

The steps of milling, chemical treatment and granulation are well understood and a number of alternative processes are available. To make production more cost-effective, more scalable processing routes are needed. For example, spray dryers are often only economical when operating at full capacity. Low capacity utilization can make the production of ceramics uneconomical. Therefore, processing concepts with low initial investment and scalable production capacity would help to solve the capacity adjustment problem.

Even though the raw materials base for silicon carbide powder is secure and inexpensive, some of the other ingredients used in the final ceramic products are critical. Rare earth oxides are used as common sintering aids, and some of the organic additives used in the ceramic feedstock are subject to availability problems. The need to secure raw materials availability is a topic that has entered public awareness in recent years.

FORMING

The heat exchanger and flow reactor plates have complex shapes and structures. This makes it necessary to employ high-quality forming and green machining.

Forming and green machining, like powder processing, are well-established ceramic processing steps, and a number of different processes are available. The process used to generate the green part is chosen according to the following criteria:

- Lot size
- Geometry
- Tolerance

- Unit production costs

Each of the forming processes used in combination with green machining has its advantages and disadvantages in respect to these parameters. Generally, production costs represent the most important criterion.

However, with increasing size and complexity, some of the established forming and green machining techniques reach their limits. New techniques for the production of ceramics, such as rapid prototyping, laser (-assisted) machining or water-jet cutting are being developed, but most of them are not yet ready for industrial-scale ceramic production.

For the design of ceramic parts, it is important to take the limits of ceramic manufacturing into account. The tolerances and geometries are generally different from those used with metals and polymers. Extending the limits of ceramic design is a task that includes all the steps along the ceramic processing route.

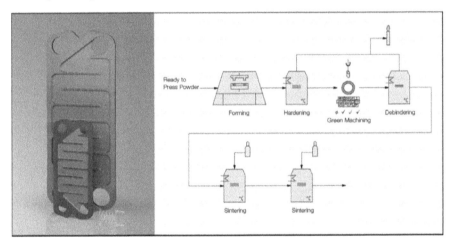

Figure 3. Overview for the processing steps of forming and sintering (source ESK).

SINTERING

Sintering is the defining step for a ceramic. It is at this stage that a lump of powder becomes a technical ceramic. Sintering of complex shapes, such as heat-exchanger and flow reactor plates (see Figure 3), requires excellent control of the sintering step and all prior processes. Once this condition has been satisfied, complex geometries can be produced economically.

After sintering, the individual plates are stacked according to standard or customized designs and subsequently diffusion-bonded to form a monolithic SiC apparatus (Figure 4 - left). After joining, no traces of the individual plates are visible (Figure 4 – right). This ensures that they have the same chemical stability and heat conductivity as other EKasic® products. The monolithic design also provides for good thermal cycling stability.

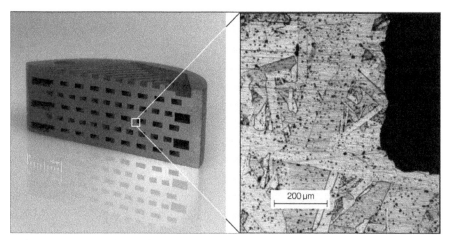

Figure 4. Seamless join of individual SiC plates to form a complex 3D microstructure (source ESK).

Sintering is a very well understood process, and models are available to describe most of the technical phenomena. Also, multiple sintering techniques, such as vacuum sintering, gas pressure sintering or hot pressing, are well-established production techniques, and equipment is available on all size scales. Nevertheless, sintering is an energy-intensive processing step. Alternative sintering techniques, such as field-assisted sintering, microwave sintering or laser sintering, have still not seen a general breakthrough, although some of these alternative techniques have gained a market share in some segments (e.g. field-assisted sintering for sintered carbide cutting tools). Reaction sintering, although established in some areas, still has its limitations.

After sintering, most critical parts require hard machining to achieve specific surface conditions (e.g. removal of the sinter skin), or precision or complex shapes. Hard machining is still the cost-defining step for most ceramic parts, and a dramatic cost reduction in this area could make ceramics attractive for many more applications.

FINAL ASSEMBLY

The stacks forming the apparatus in Figure 5 are either standardized or designed individually. A series of stacks is combined to form a flow reactor including e.g. modules for mixing, residence or quenching. System integration includes the steps of:
- Stacking SiC modules in the desired sequence within a stainless-steel frame
- Installing polymer interconnector plates with minimum dead volume for contacting and sealing the monolithic modules properly
- Attaching Swagelok fittings for the product inlets and outlets

Assembly expertise is critical for the market success of ceramic systems. It allows ceramic manufacturers like us to become system manufacturers or system integrators.

Generally, the joints of ceramics and non-ceramics are critical, and research is still progressing in this field. From a production standpoint, assembly has to be integrated into the ceramic processing workflow.

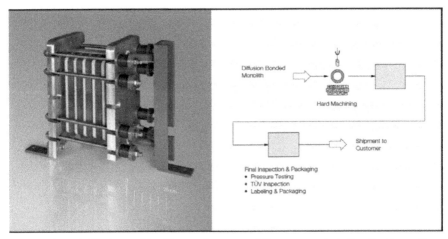

Figure 5. Final assembly of heat exchangers and flow reactors including final inspection and packaging (source ESK).

SUMMARY

With the example of heat exchangers and flow reactors, the current state of development and typical challenges for each of the process steps have been described from our perspective.

The property fingerprint for heat exchangers and flow reactors includes high thermal conductivity, abrasion resistance and chemical stability. EKasic® sintered silicon carbide offers this combination of properties and is the material of choice. By producing not just the ceramic part but the full system, companies like us are developing from a ceramic manufacturer into a system manufacturer.

REFERENCES

[1]J. Eichler, Industrial Applications of Si-based Ceramics, *Journal of the Korean Ceramic Society* **49** [6] 561-565, 2012

[2]A.W. Weimer, Carbide, Nitride and Boride Materials Synthesis and Processing, *Chapman & Hall,* 1997

[3]H.B. Ries, Herstellung von Siliciumcarbid-Körnungen, *Keramische Zeitschrift* **606** [9/10], 2004

[4]W. Kollenberg, Technische Keramik – Grundlagen, Werkstoffe, Verfahrenstechnik, *Vulkan Verlag,* 2009

[5]Source: www.alibaba.com

FABRICATING SUCCESSFUL CERAMIC COMPONENTS USING DEVELOPMENT CARRIER SYSTEMS

Tom Standring[a], Bhupa Prajapati[b], Alex Cendrowicz[b], Paul Wilson[c] and Prof. Stuart Blackburn[a]
a, School of Chemical Engineering, University of Birmingham, B15 2TT. b, Ross Ceramics Ltd, Denby, DE5 8NA. c, Rolls-Royce Plc, Derby, DE24 8BJ.

ABSTRACT

Fabrication of complex ceramic components can be achieved through the process of Particle Injection Moulding (PIM). In this process a feedstock comprising of ceramic powders suspended in an organic carrier system (binder) is forced, at elevated temperatures, into preformed moulds. The organic binder is then thermally removed and the remaining powders are sintered to consolidate the ceramic. The carrier system is critical for developing the desired properties of the feedstock to enable flow and mould filling during injection and to sustain the integrity of the formed component during the subsequent firing process.

This investigation describes the effect of three development carrier systems of high, medium and low viscosity, containing identical components but differing ratios, on component fabrication. The feedstock was optimised in solids loading and plasticiser additions. A PIM press was used for the moulding and before injection the feedstock was conditioned for both short and long mixing times. The process was characterised through rheometry, mechanical testing and interferometry.

Experimental data illustrates that conditioning time and viscosity of the binder affects the surface roughness of components. Conditioning time also influences the flexural stress profile of the components under load. The ceramic strength properties using the low and medium viscosity binders were similar, however, a reduction in strength properties was observed when using the high viscosity binder.

INTRODUCTION

Complex ceramic components are often fabricated by injection moulding due to the ability of the process to allow high throughput and precision components to be formed (1). Component formation involves the injection of feedstock at elevated temperatures (ceramic paste) into pre-formed moulds, with the feedstock constituents being the desired ceramic powders and an organic carrier system (binder). Once the paste is successfully moulded, the resulting green component is fired to remove the carrier system before sintering the remaining powders, to develop the required fired component properties.

The fired ceramic components are used in a secondary process which involves interactions with molten metal. Here the component must be chemically compatible with the metal, be crushable when the metal solidifies, as well as maintaining enough strength for handling purposes and to be resistant to thermal shock and dimensional changes. Components also need to yield a suitable surface finish and be removed from the final metal component without damage (2).

The carrier system has important influences on developing a number of the important criteria required of the fired components with often the main differences in ceramic injection moulding systems directly related to the carrier system, highlighting its importance in the process (3). The following study investigates the success of three development carrier systems on simple component formation, with the study focussing on rheology, mechanical properties and surface roughness. The effect of feedstock conditioning before injection is also described.

EXPERIMENTAL TECHNIQUES

Materials

Each of the three development binders was blended in a planetary mixer at 125°C. The binders were paraffin wax based with different amounts of polymer additives incorporated to modify the properties. The ceramic powder mix used was silica based and was a proprietary formulation supplied by Ross Ceramics Ltd. (RCL). Stearic acid was utilised as the wetting agent in the investigations, and this is a common material used for the role in binder formulations.

Binder Viscosity

Binder viscosity was determined on a Malvern Gemini HR Nano controlled stress rheometer. The temperature effects on viscosity for the three development binders were determined using a 40 mm parallel plate and constant shear rate of 100 s^{-1}, allowing a direct comparison with literature data describing carrier systems used in ceramic injection moulding (CIM) applications, with a 150 µm gap between the plates used (3). Samples were cooled from a temperature of 95°C down to 40°C at a rate of 10°C.min^{-1}.

Feedstock Preparation

The critical loading for each binder and powder formulation was determined experimentally. Numerous techniques have been developed and used to attempt to describe this critical loading, with common techniques including rheometry measurement. A more novel method involves using density measurements as described in a paper by Shengjie et al. (4). For the three development binders the critical loading for each was determined by this pycnometry method, using a 0.001 m^3 mixer. Once the critical loading was determined a larger sample was formulated in a 0.01 m^3 mixer, at two volume percent solids below the pre-determined critical loading, for use in the component moulding trials. Feedstock preparation involved adding pre-blended powder and stearic acid to the molten binder, before using the RCL mixing parameters to manufacture the feedstock.

The equivalent amount for '5 layers' of stearic acid adsorption onto the powder surface was used. The 'layers' of absorption was a theoretical value based on the amount of plasticiser that could adsorb based on 100% efficiency. A monolayer was calculated by finding the surface area of powders by Brunauer-Emmett-Teller (BET) analysis. The amount of surfactant required to cover the powder surface was determined by finding the projected surface area of one surfactant molecule and dividing this by the total powder surface area to get the total number of surfactant molecules required. The projected surface area of surfactant on the powder surface (assumed to be spherical) was determined by using the diameter of the stearic acid polar head as 5.11×10^{-10} m as stated by Li et al. (5). By using Avogadro's number and the molar mass of stearic acid, the mass of the surfactant required to form a monolayer could be determined.

Feedstock Rheology

The Malvern Gemini HR Nano controlled stress rheometer was also utilised for dynamic rheological measurements, to determine the viscoelastic nature of the injected pastes with respect to temperature changes. This was conducted using temperature sweeps in a pre-determined linear viscoelastic region (LVR).

The LVR was measured by dynamic testing performed using a fixed frequency (1 Hz) at 80°C. The amplitude of the applied stress was increased stepwise. The LVR denotes the range in which the elastic modulus, G', remained constant. Above a critical stress, G' begins to decrease and this region with a non-linear dependence results in the describing equations not being appropriate. For testing a parallel plate geometry was used, incorporating a 25 mm serrated plate to limit paste slip, and a 500 μm gap. Temperature sweeps were then conducted in the LVR region using the same geometry and a temperature ramp rate of $10°C.min^{-1}$.

Each feedstock was also characterised by a Rosand precision advanced capillary extrusion rheometer as the test method offers similar flow conditions to those imposed on the paste during injection into a mould. Feedstock was tested at the injection temperature of 80°C under shear rates of 100, 250, 500, 750, 1000, 2500, 4000, 5000 and 6000 s^{-1}. The top end shear rate was determined by equipment limitations.

Component Fabrication

Each feedstock was conditioned before injection for both short and long mixing times. After conditioning each feedstock was injected at 80°C into test bar moulds (dimensions in mm, 100x12x4). Half the test bars were tested in the as moulded (green) state, and half were fired prior to testing using a proprietary RCL firing cycle.

Mechanical Testing

Mechanical testing for both green and fired components was conducted on a Lloyd Instruments load frame. The three-point bend test was used for the testing, using a span of 80 mm and cross head speed of 10 $mm.min^{-1}$. The flexural testing enables the flexural modulus, flexural strength (otherwise known as modulus of rupture, MoR) and energy to fracture to be determined. The flexural modulus is the gradient of the flexural stress versus flexural strain graph, with flexural strength determined by the point of fracture. The energy of fracture is described by the area encompassed under this graph up to the point of fracture.

Porosity Measurement

The Archimedes water absorption test was used for determining the porosity of the fired components by means of water infiltration. Samples were suspended in boiling water for two hours, and samples were weighed before and after. The apparent porosity can be determined, using equation 1, using the mass of a dry, immersed and infiltrated test piece in water (6). In equation 1, π_a is the apparent porosity (%), m_1 is the mass of a dry test piece (grams), m_2 is the apparent mass of immersed test piece (grams) and m_3 is the mass of infiltrated test piece (grams).

$$\pi_a = \frac{m_3 - m_1}{m_3 - m_2} \cdot 100 \tag{1}$$

Surface Analysis

Surface roughness of components was conducted using an interferometer. Each component was tested at three points along its length. The software processes the data to form 2D images, 3D images or histograms to describe the variation in roughness on the surface. 2D images have been used for this investigation with each image displaying four numerical values to describe the roughness (7). The Sa value is the arithmetic average, Sq is the root mean square, Sku describes the data skew from a normal distribution and Sy describes the difference between the highest peak and the lowest trough. For statistical measurements the root mean square is often the preferred measure.

RESULTS AND DISCUSSION

Binder Viscosity
There is a clear viscosity difference between the three binders, highlighted in Figure 1. The binders have been compared against paraffin wax for reference. Literature states that the binder melt viscosity should be below around 10 Pa.s at the moulding temperature, which encompasses all three development binders (3). The melt region of the binders can be described through the Arrhenius equation (8). The increase in melt viscosity is attributed to the longer chains found in the additives causing increasing interlinking and hence more resistance to flow.

As the binders are cooled from melt they solidify which is described by the steep viscosity increase in Figure 1. An interesting observation is shown for binder 3 were the onset of viscosity increase differs from binder 1 and 2 even though the same components are used in the make-up. The binders solidify in one region denoting the binder components being compatible with one another. At the injection temperature of 80°C, paraffin wax, binder 1, binder 2 and binder 3 have the respective viscosities of 0.004, 0.011, 0.074 and 0.7 Pa.s.

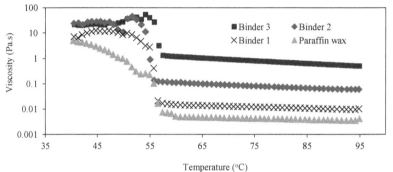

Figure 1: Effect of temperature on viscosity of the three development binders
(Shear Rate of 100 s^{-1} at a ramp rate of 10°C.min^{-1}).

Optimised Solids Loadings
Determination of the critical loading for a set powder and binder combination offers valuable information into the processing window that should be used to enable successful components to be fabricated. The critical loading is denoted at the stage whereby the 'binder' completely fills the voids between the powder particles, at this point the viscosity is infinite and no flow will occur. The carrier system achieves the desired fluidity by filling the space between the powder particles above that required to fill the voids, ideally to the exclusion of air, and this is done with sufficient excess to allow the particles to rotate in shear or extensional flow. The degree of this excess can have large implications on the flow properties of the paste as well as on mass and heat transfer and on defect formation. It has been proposed in literature that a window exists which is ideal for injection moulding applications, with this lying 2 to 5 volume percent below the critical loading (3).

The result of the experiments to determine the critical loadings of feedstock 1, 2 and 3 are displayed in Figure 2, 3 and 4, respectively. The feedstock number describes the binder which the feedstock was formed from so; feedstock 1 denotes the feedstock formed from binder 1. The critical loading is denoted at point 1 were there is just sufficient binder to fill all voids between particles, leaving no excess binder to allow flow.

As loading is increased beyond this point a reduction in density is observed as enclosed air inclusions start to form in the voids between particles, due to there no longer being enough binder to fill all the voids.

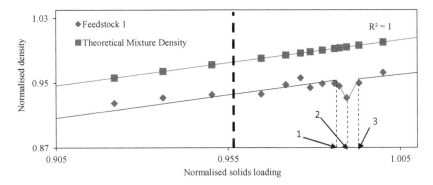

Figure 2: Experimental testing to determine the critical loading for Feedstock 1.

Density reduction continues to a minimum at point 2 were the air inclusions grow to a maximum size with increasing solid loadings. There is then a subsequent increase in density as the air pockets become so large they are no longer encapsulated. From point 3 there is no longer any encapsulated air and the true density of the material is observed. It can be observed that the theoretical density is greater than the feedstock density which will be down to the errors in measurements.

Feedstock 1 and 2 achieve very similar normalised critical solids loadings, 0.987 and 0.984 respectively. However, a considerable drop to 0.974 is observed for feedstock 3. This phenomenon could be attributed to the greater viscosity of binder 3 being unable to completely fill the voids between particles resulting in air voids forming at lower solids loadings. The dotted lines highlight the solids loading used for injection of each feedstock, which is set at two volume percent below point 1 from the pycnometer data.

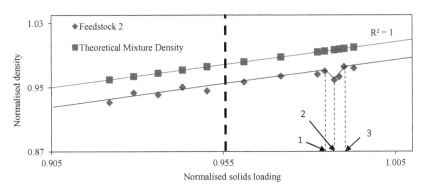

Figure 3: Experimental testing to determine the critical loading for Feedstock 2.

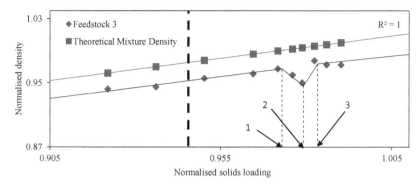

Figure 4: Experimental testing to determine the critical loading for Feedstock 3.

Paste Rheology

A general guideline for suspensions states that the viscosity at the lower shear rate end (100 s^{-1}) should not exceed 1000 Pa.s, and this has been successful for the development of a number of compositions (9). The viscosity of the development ceramic pastes by capillary rheometry is below this 1000 Pa.s, with profiles for each shown in Figure 5.

It can be shown that all three pastes experience shear thinning character with increasing shear rate, which is an attractive behaviour as it gives the pastes the best chance of filling the moulds. It is interesting to note that feedstock 2 has the lowest viscosity at low shear rates but as shear rate is increased beyond 1000 s^{-1} the viscosities of all three pastes become comparable. In a typical injection moulding process the shear rate ranges between 100 to 1000 s^{-1} (10), however, from around 5000 s^{-1} it appears that feedstock 1 and 2 may experience a degree of phase migration due to the extreme drop in viscosity.

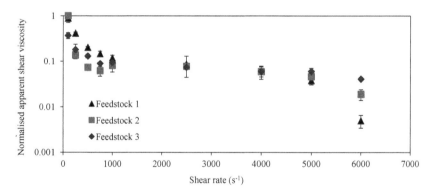

Figure 5: Effect of shear rate on feedstock viscosity at injection temperature using capillary rheometry (80°C).

Describing the injected ceramic paste by dynamic testing supplies information into how the viscoelastic paste properties changes during the injection process. To determine the effect of temperature on the pastes data must be collected in the linear viscoelastic region (LVR), for the development pastes this is found from Figure 6. The dotted lines enclose a region were all three pastes are within a LVR. It is observed that feedstock 1 has the smallest LVR which can be put down to its low viscosity binder and the resulting weak structure this gives to the paste. A serrated plate was used for testing to try and reduce slip; however, fluctuating data points still appear to be present at low strains possibly describing slip or simply noise.

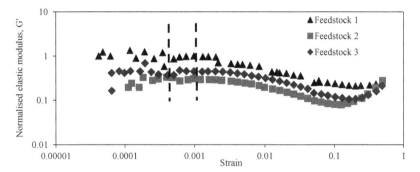

Figure 6: Strain sweep test (at 1 Hz) for feedstocks at 80 °C.

The change in each feedstock properties with decreasing temperature has been characterised in Figure 7. The phase angle gives information regarding the phase state of the ceramic paste as temperature is reduced. A viscous liquid would have a phase angle of ninety degrees whereas an elastic solid-like material would have a phase angle of zero degrees. At injection the phase angle of pastes 1, 2 and 3 were 29, 48 and 60 respectively. Feedstock 1 exhibits predominantly elastic behaviour, whereas feedstock 2 would be described as being neither elastic nor viscous dominated with feedstock 3 being slightly more liquid like. The phase angle of each paste remains consistent until binder solidification occurs which coincides with a reduction in the paste phase angle. From about 48°C the phase angle of all pastes is around zero, which describes solid-like materials.

The stiffness of a paste is characterised by the complex modulus, which is a factor of the elastic and viscous modulus that describe the solid and liquid elements of the material. An increased complex modulus details a stiffer paste. As temperature decreases and the binder solidifies in the paste there is an associated increase in the complex modulus.

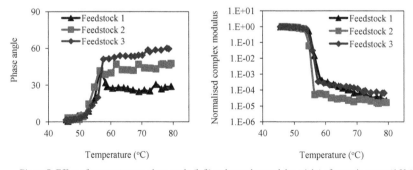

Figure 7: Effect of temperature on phase angle (left) and complex modulus (right) of ceramic pastes (1 Hz).

Mechanical Properties of Fabricated Components

All three pastes were successfully moulded into the test bars at 80°C, using the injection press, when both short and long conditioning times were utilised. It should be noted that component 1 has been formed using feedstock 1, component 2 from feedstock 2 and component 3 from feedstock 3.

For green components the length of conditioning before injection appears to have a minimal affect on the component behaviour in flexural testing. The representative profiles, Figure 8, illustrate that a more viscous binder results in a reduced flexural modulus of the component, which is thought to be due to the increased interlinking in more viscous binders developing more flexible properties in the component. The box plots in Figure 9 shows that all three materials exhibit similar flexural strengths in the green state.

Box plots are forming using the interquartile range (IQR) of the data set. The top and bottom of the shaded box denote the upper (Q3) and lower quartiles (Q1) with the centre line describing the median (Q2). Whiskers (vertical line) extend from the top and bottom of the box to the highest and lowest data points which are within Q1 - 1.5 (IQR) or Q3 + 1.5 (IQR). Any asterisks show points (outliers) outside of these regions.

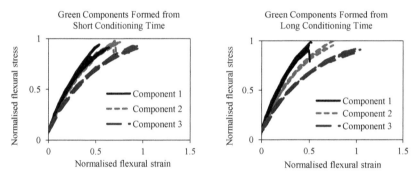

Figure 8: Profile traces describing the effect of an applied load to green components, conditioned for a short and long mixing time.

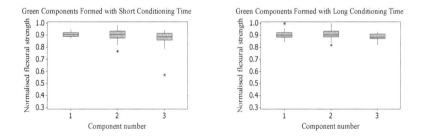

Figure 9: Flexural strengths of green components from short conditioning time (left) and long mixing time (right).

For the fired components the effect of the applied load on fracture failure is shown in Figure 10. Increasing the length of conditioning appears to increase the fracture strength of components, observed more clearly in the box plots of Figure 11. Another interesting observation is that component 2 has the greatest flexural strength with component 1 slightly lower; however, component 3 has significantly lower flexural strength as well as a lower flexural modulus.

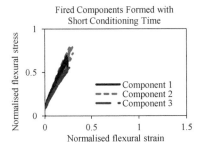 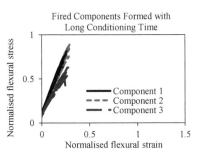

Figure 10: Profile traces describing the effect of an applied load on fired components, initially conditioned for short (left) and long (right) conditioning time.

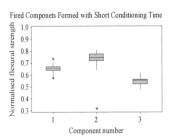 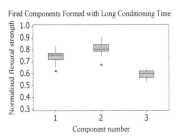

Figure 11 Flexural strength for fired components from short conditioning time (left) and long conditioning time (right).

There appears to be a tentative trend, Figure 12, between the flexural strength and the apparent porosity. The highest flexural strength found in component 2 coincides with the lowest porosity and the lowest flexural strength of component 3 is linked to the highest porosity. However, there is likely to be errors in measurements especially as the measurement technique does not register enclosed porosity which would have a large effect on component strength.

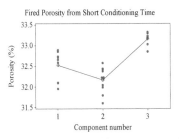 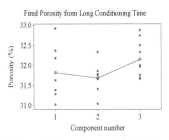

Figure 12: Apparent porosity of fired components using Archimedes method from short conditioning time (left) and long conditioning time (right).

Surface Roughness

It is important that fired components possess an adequate surface finish to minimise defects to cast metal in the secondary process steps. The surface analysis of the green components are shown in Figure 13, with the scale on the right hand side of the images highlighting the relative difference in roughness across the sample.

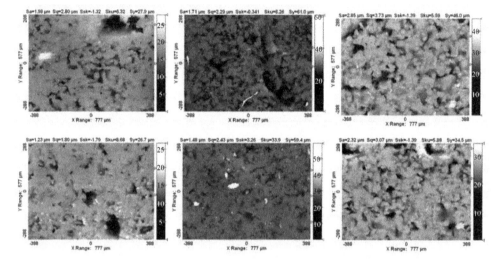

Figure 13: Green component roughness. Left to right Component 1-3, with the top line being short conditioning time and the bottom line being long conditioning time.

There is an observed change in the surface of the different components with a particularly marked difference noticeable between component 1 and 3. This suggests the viscosity of the binder affects the complete filling of a mould. Conditioning time from the images does not appear to have any significant effect on surface roughness. However, by plotting the Sq values for green component roughness it does appear that Sq reduces with longer conditioning times, Figure 14. This may be due to longer conditioning allowing slightly improved wetting of powders allowing a better dispersion to be achieved.

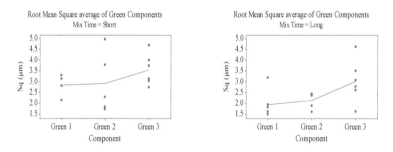

Figure 14: Surface roughness of green components formed from short (left) and long (right) conditioning times.

The surface images of fired component are illustrated in Figure 15, were the fired components have a rougher surface finish than the green components which can be explained due to the binder removal process. Data shows the conditioning time used in forming green components does not affect the final fired surface roughness; as both short and long conditioning have comparable roughness values.

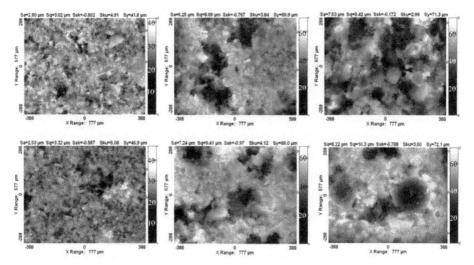

Figure 15: Fired component roughness. Left to right Component 1 -3, with the top line being short conditioning time and the bottom line being long conditioning time.

There is a clear observable difference between the roughness of component 1 and components 2 and 3, which is confirmed in Figure 16 and appears to show a trend between component number and surface roughness. This suggests that certain constituents in the more viscous binder are linked to this increase in roughness.

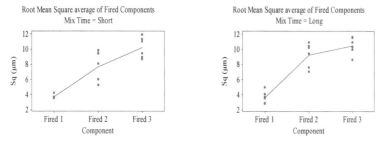

Figure 16: Surface roughness of fired components formed form short and long conditioning times.

CONCLUSIONS

Feedstock produced from a lower viscosity binder exhibits a greater overall shear thinning character.

For green components it was noted that the conditioning time results in no observable difference to the strength properties, however, an increase in binder viscosity enables greater flexural strain before component fracture.

For fired components, the conditioning time did not affect the strength properties when using the same binder. The ceramic strength properties of the fired components appears to be linked to the apparent porosity, with the data suggesting an optimum strength can be found at a binder viscosity between that of binder 1 and 3.

Data suggests there are correlations linking an increasing viscosity of the binder used in the feedstock to increasing surface roughness of both green and fired components. It was also observed that long conditioning times appeared to reduce surface roughness in green components but that conditioning time has no observable effect on the final fired component roughness.

ACKNOWLEDGMENT

The author would like to acknowledge with thanks the financial support received from the EPSRC and Rolls-Royce plc.

REFERENCE LIST

(1) Z.Y.Lui, N.H.Loh, S.B.Tor, K.A.Khor, Y.Murakushi, R.Maeda. Binder System for Micropowder Injection Moulding. Materials Letters 2001;(48):31-8.
(2) A.A.Wereszczak, K.Breder, M.K.Ferber, T.P.Kirkland, E.A.Payzant, C.J.Rawn, et al. Dimensional changes and creep of silica core ceramic used in investment casting of superalloys. Journal of Material Science 2002;37:4235-45.
(3) B.C.Mudsuddy and R.G.Ford. Ceramic Injection Moulding. Chapman and Hall; 1995.
(4) Y.Shengjie, Q.F.Li, M.S.Yong. Method for determination of critical powder loading for powder-binder processing. Institute of Materials, Minerals and Mining 2006;49(3):219-23.
(5) Y.Li, X.Lui, F.Luo, J.Yue. Effects of surfactant on properties of MIM feedstock. Transactions of Nonferrous Metals Society of China 2007;17:1-8.
(6) European Committe for Standardisation. Methods of test for dense shaped refractory products: Part 1 Determination of bulk density, apparent porosity and true porosity. 1995.
(7) Mitutoyo. Surface Roughness Measurement. Mitutoyo American Corporation; 2009. Report No.: Bulletin No. 1984.
(8) C.K.Schoff and P.Kamarchik. Rheology and rheological Measurements. 2005.
(9) R.W.Cahn, P.Haasen, E.J.Kramer. A comprehesive treatment; Processing of ceramics. VCH; 1996.
(10) R.M.German and A.Bose. Injection Molding of Metals and Ceramics. 1997.

OPTIMIZED SHAPING PROCESS FOR TRANSPARENT SPINEL CERAMIC

Alfred Kaiser[1], Thomas Hutzler[2], Andreas Krell[2], and Robert Kremer[3]
[1]LAEIS GmbH, Wecker, Luxembourg
[2]Fraunhofer Institute for Ceramic Technologies and Systems (IKTS), Dresden, Germany
[3]ALPHA CERAMICS GmbH, Aachen, Germany

ABSTRACT

Bulky transparent ceramic, especially spinel ($MgAl_2O_4$), can be used for applications such as high-energy laser windows and lightweight armor. One of the traditional routes to manufacture transparent spinel plates includes the steps of material preparation, uniaxial pressing, cold isostatic pressing (CIP), de-bindering and hot isostatic pressing (HIP). When larger sizes are required, CIP can become one of the bottlenecks of the process chain. The paper shows, how an optimized uniaxial hydraulic pressing process with an evacuated mould allows to avoid the cold isostatic pressing completely. The process is described and the first results of the investigation of transparent spinel properties are discussed. This simplified process will allow to reduce the manufacturing costs for larger sized transparent spinel significantly and/or improve the production capacity.

INTRODUCTION

Spinels are a group of minerals of general formulation AB_2O_4 (where A is a bivalent and B is a trivalent cation), which crystallise in the cubic crystal system with A and B occupying some or all of the octahedral and tetrahedral sites in the lattice. Spinel ($MgAl_2O_4$), after which the spinel group is named, in its pure form (single crystal) is a colourless, transparent material with high hardness and excellent transmission from the ultraviolet (0.2 µm) to the mid-infrared (5 µm) region. This makes spinel an interesting material for numerous applications such as high-energy laser windows and lightweight armor[1]. However, single crystal spinel is difficult to make in dimensions greater than a few millimeters using traditional high temperature (>2000 °C) melt growth techniques[2]. Various approaches to transparent polycrystalline spinel have been made, e.g. by using sintering aids like LiF or using sub-µm spinel powder[3-8] and applying different shaping and sintering technologies[9, 10]. One of the traditional routes to manufacture polycrystalline transparent spinel plates includes the steps of material preparation, uniaxial pre-pressing, cold isostatic pressing (CIP), de-bindering and hot isostatic pressing (HIP). When larger sizes are required, CIP can become one of the bottlenecks of the process chain.

Aim of this work was to investigate a more economic route to prepare large-sized transparent spinel plates by optimization of the uniaxial pressing process and eliminating the cold isostatic pressing process completely.

MATERIAL PREPARATION

Commercially available high purity spinel powder, especially developed for transparent ceramics applications, was prepared by IKTS for the tests (see Table I).

Table I. Technical data of spinel powder used as starting material.

supplier	BAIKOWSKI, France
product code	S30CR
BET specific surface area	30 ± 5 m²/g
d_{50} (PSD Sedigraph)	0.2 µm
crystalline phase (XRD)	≥ 99 % spinel

The aggregated raw powder (Fig. 1 left) with high specific surface area (Table I) and a majority of particles ranging between 60 and 90 nm (Fig.1 right), was deagglomerated by milling in a NETZSCH horizontal disk mill type LME 4 for one hour with deionized water and with about 7 % of organic binders.

Figure 1.   SEM micrographs of S30CR powder as supplied

After milling the slurry was dried in a freeze dryer type EPSILON 2-45D (supplier: CHRIST, Germany). For obtaining ready-to-press granulates the very fluffy dried spinel powder was screened with a mesh size of 250 μm. Figure 2 shows a SEM photograph of the powder as it was used for the pressing trials. Despite their flaky structure the agglomerates show a sufficient flowability for even mould cavity filling.

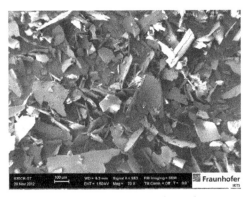

Figure 2.   SEM micrograph of processed S30CR powder ready to press

UNIAXIAL HYDRAULIC PRESSING
For the uniaxial pressing trials a LAEIS hydraulic press type ALPHA 1500/120 was used (Figure 3), which is available at the LAEIS technical center ALPHA CERAMICS in Aachen. Features of this type of uniaxial hydraulic presses, available sizes and application examples for shaping of advanced ceramic products are described elsewhere[11-13].

Figure 3.   Hydraulic press LAEIS ALPHA 1500-120

The main technical data are summarized in Table II. The press was equipped with a vacuum unit to allow the plates being pressed in an evacuated mould. Thus, a major part of the air is removed before compaction, which helps to avoid delamination or other pressing defects due to entrapped air[14].

Table II. Technical data of LAEIS hydraulic press ALPHA 1500/120.

press type		ALPHA 1500/120
pressing force	kN / t	15,000 / approx. 1,500
useful die area	mm x mm	1,320 x 640
maximum filling depth	mm	120
maximum ejection force	kN	300

The first tests were performed in a mould with the dimensions 191 x 142 mm². The mould was manually filled with a filling depth of 25 mm and pressing under vacuum took place at pressures between 100 and 250 MPa. The resulting green plates showed high strength with sharp contours and could be handled very easily. A green density of up to 1.9 g/cm³ was reached. Depending on the applied pressure, a densification factor between 3.5 and 4.0 was obtained. Figure 4 shows that with higher pressures the density could still be increased significantly, however the mould design capacity was limited to 250 MPa. Some of the plates were repressed in a cold isostatic press at 350 MPa, ending up with a green density of approximately 1.97 g/cm³, which is similar to what can be extrapolated from the uniaxial pressing values.

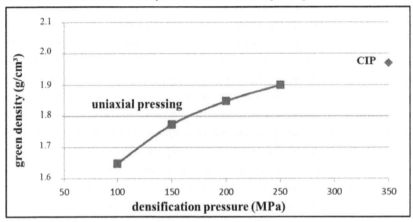

Figure 4.   Green density of spinel plates 191 x 142 mm² pressed with uniaxial press (□)
            and with cold isostatic press (◊) at various pressures

A second test series was made on the ALPHA 1500/120 press using a mould with the dimensions 300 x 400 mm². Again the mould was filled manually with a filling depth of 50 mm and pressed under vacuum (see Figure 5). Due to the large pressing area the maximum applicable pressure in this series was 120 MPa. The green density of the large plates pressed with 120 MPa was between 1.70 and 1.71 g/cm³, which fits perfectly with the values obtained for the smaller plates (fig. 4). The densification factor was approximately 3.5 (i.e. green plate thickness of approximately 14 mm), again matching the corresponding results of the smaller plates. Also these large plates showed good strength and well defined contours and could be handled and shipped safely.

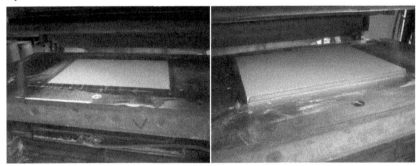

Figure 5.   Volumetrically filled mould 300 x 400 mm² and green plate after ejection

These tests show that scaling up works and that it is possible to press also large-sized spinel plates without defects to a sufficient green density. If a larger press is used (presses are available up to a pressing force of 64,000 kN = 6,400 t) and the mould is designed accordingly, pressures up to approximately 350 or 400 MPa can be realized. Therefore it would be possible to press plates e.g. with green dimensions of 400 x 400 mm² to similar green densities which can be achieved by cold isostatic pressing.

THERMAL PROCESSING

All activities regarding thermal treatment of the green plates were conducted at the IKTS in Dresden. Removal of organic components (debindering) was performed at 800 °C in air. To avoid thermally induced stress, a low heating rate of 6 K/h was used. After debindering, the smaller samples were pre-sintered without pressure up to closed porosity (> 95%...98% relative density) at temperatures between 1550 and 1590 °C for 2 hours, again with low heating and cooling rates. The necessary temperature for a sufficient sintering process (elimination of the open porosity) was approx. 35 K higher for the uniaxially pressed plates compared to the CIP repressed plates, apparently due to differences in the density and in the homogeneity of particle coordination in these differently shaped bodies.

The pre-sintered plates were hot isostatic pressed in argon at 1590 °C for 15 hours and at a maximum pressure of 185 MPa. The HIP-process was performed in a hot isostatic press with graphite heaters and a useful volume of 32 dm³ (supplier: EPSI, Belgium). After HIP, all plates were at the same density of 3.58 g/cm³ and showed transparency. Solely the plates which were pressed with only 100 MPa had some opaque areas where the porosity was not eliminated completely. Figure 6 shows some photographs of samples prepared from the smaller plates after polishing.

Figure 6.   Transparent spinel samples prepared from 191 x 142 mm² plates
(thickness approx. 4 mm)

The large plates were pre-sintered at 1600 °C for 2 hours and hot isostatic pressed at 1600 °C for 15 hours. Plates with a density of 3.581 g/cm³ were obtained, showing good transparency except very few small areas close to the surface (see Figure 7). The overall linear shrinkage in length and width amounts to approximately 23 %, the shrinkage in thickness is even higher (27 %).

Figure 7.   Large spinel plate after sintering (left) and transparent after HIP + both
sides ground and polished (right)

SAMPLE CHARACTERIZATION

Samples with approximately 20 mm diameter and 4 mm thickness were prepared from the
smaller plates as well as from the large plates. They were ground and polished and the real-inline-
transmission (RIT) was determined. The real in-line transmission was measured at 640 nm wave
length with an aperture of about 0.5° (LCRT 2000, Gigahertz Optics, Puchheim, Germany). This
measurement excludes scattered amounts which are included in "in-line" data of commercial
spectrometers with larger effective apertures of 3-5° [15,16]. The results are shown in table III.

Table III. Transmission values of different spinel samples.

plate size (after HIP)	pressure		RIT	thickness
mm x mm	MPa		%	mm
110 x 130	100		78.8	3.94
110 x 130	150		79.3	3.94
110 x 130	150	+ 350 (CIP)	81.6	3.94
110 x 130	200		80.3	3.58
110 x 130	200	+ 350 (CIP)	81.4	3.94
110 x 130	250		81.4	3.87
220 x 300	120		76.1	3.94
220 x 300[*]	120		61.5 – 72.8 (69.6 – 77.1)[**]	5.95

[*] toll polishing in an industrial plant; surface roughness Rz = 0.17 µm compared to
Rz = 0.03 µm of the laboratory prepared samples
[**] recalculated to thickness = 4.0 mm

The RIT values show a comparable high fluctuation, when measured in a regular grid over the whole area of the plate. There is also a number of macroscopic visible defects (pores, contaminations?) spread over the whole plate area, and the large-sized plates have different shades of greying despite identical production conditions. The Vickers hardness of the large plate was determined to 1277 ± 24 HV10 as can be expected[17].

Average grain sizes of the dense microstructures were determined on SEM micrographs by the linear intercept approach (average grain size =1.56 average intercept length[18]). SEM mircrographs of polished cross sections show a bimodal structure (Fig. 8): an average grain size of 2.5 µm was measured in the fine grained areas whereas grain sizes between about 25 and 110 µm were observed in the coarser parts of the microstructure.

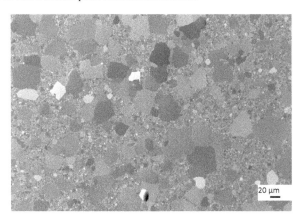

Figure 8.   Example of microstructure after HIP at 1600 °C

SUMMARY AND OUTLOOK

The results reported in this paper are based on a preliminary feasibility study. It basically proved the possibility to produce crack free transparent spinel plates, also of larger size, using vacuum assisted uniaxial hydraulic pressing technology and waiving the expensive redensification by CIP. There are still open questions remaining and additional development work need to be done, eg.:

- evaluation of a powder preparation process which is better suited for large scale production than the lab scale freeze drying process
- determination of the reason(s) for the remaining visible defects and elimination of such reasons
- increasing the pressure applied during the uniaxial shaping (adaption of mould design and/or use of larger press)
- optimization of pressing parameters, vacuum application
- optimization of debindering, sintering, and HIP parameters

The intermediate results, derived from a very small number of pressing trials and sample characterization,   provide a very good starting point for further evaluation and for the development of a successful production scale technology.

REFERENCES

[1]D.C. Harris; History of development of polycrystalline optical spinel in the U.S., *in: Window and Dome Technologies and Materials IX, Proceedings of SPIE Volume 5786, ed. Randal W. Tustison (2005)*

[2]J.S. Sanghera, G. Villalobos, W. Kim, S. Bayya, and I.D. Aggarwal; Transparent spinel ceramic, *NRL Review (2009)*

[3] A. Goldstein; Correlation between $MgAl_2O_4$-spinel structure, processing factors and functional properties of transparent parts, *J. Eur. Ceram. Soc. 32 (2012) [11] 2869-2886*

[4] R. Cook, M. Kochis, I. Reimanis, H-J. Kleebe; A new powder production route for transparent spinel windows: powder synthesis and window properties, *in:Window and Dome Technologies and Materials IX, Proceedings of SPIE Volume 5786, ed. Randal W. Tustison (2005)*

[5] G. Villalobos, J.S. Sanghera, and I.D. Aggarwal; Transparent ceramics: magnesium aluminate spinel, *NRL Review (2005)*

[6] I. E. Reimanis, K. Rozenburg, H-J. Kleebe, R. L. Cook; Fabrication of transparent spinel: the role of impurities, *in:Window and Dome Technologies and Materials IX, Proceedings of SPIE Volume 5786, ed. Randal W. Tustison (2005)*

[7]A. Krell, T. Hutzler, J. Klimke, A. Potthoff; Nano-processing for larger fine-grained windows of transparent spinel, *Proceedings of 34th International Conference on Advanced Ceramics and Composites; Daytona Beach, Jan 24-29, 2010, The American Ceramic Society; ed. J.J. Swab, S. Mathurs & T. Ohji, The American Ceramic Society; ed. M. Halbig & S. Mathur (= Ceram. Eng. Sci. Proc. 31 (2010)[5] 167-182).*

[8]A. Krell, T. Hutzler, J. Klimke, A. Potthoff; Fine-grained transparent spinel windows by the processing of different nanopowders, *J. Am. Ceram. Soc. 93 (2010) [9] 2656-2666*

[9]A. LaRoche, K. Rozenburg, J. Voyles, L. Fehrenbacher, G. Gilde; An economic comparison of hot pressing vs. pressureless sintering for transparent spinel armor, *in: Advances in Ceramic Armor IV: Ceramic Engineering and Science Proceedings 29 (2009) [6] (ed. L. P. Franks), John Wiley & Sons, Inc., Hoboken, NJ, USA*

[10]K. Morita, B.-N. Kim, K. Hiraga, H. Yoshida; Fabrication of high-strength transparent $MgAl_2O_4$ spinel polycrystals by optimizing spark-plasma-sintering conditions, *Journal of Materials Research 24 (2009) [9] 2863-2872*

[11]A. Kaiser, R. Lutz; Uniaxial hydraulic pressing as shaping technology for advanced ceramic products of larger size, *Interceram 60 (2011) [3] 230-234*

[12]R. Lutz; Use of closed loop controls in hydraulic press forming of ceramic products to obtain highest dimensional accuracy, *Proceedings of the International Colloquium on Refractories, Aachen (2004) 222-224*

[13]A. Kaiser; Shaping of large-sized sputtering targets, *Proceedings of 36th International Conference on Advanced Ceramics and Composites; Daytona Beach, Jan 22-27, 2012 The American Ceramic Society; ed. M. Halbig & S. Mathur*

[14]A. Kaiser, R. Kremer; Fast acting vacuum device: guaranteed quality for pressed refractories, *Interceram Refractories Manual (2003) 28-33*

[15]R. Apetz, M.P.B. van Bruggen; Transparent alumina: a light scattering model, *J. Am. Ceram. Soc. 86 (2003) [3] 480-486*

[16]H. Yamamoto, T. Mitsuoko, S. Iio; Translucent polycrystalline ceramic and method for making same, *Europ. Patent Application EP - 1 053 983 A2, IPK7 C04B35/115, 22.11 (2000)*

[17]A. Krell, A. Bales; Grain size-dependent hardness of transparent magnesium aluminate spinel, *Int. J. Appl. Ceram. Technol. 8 (2011) [5] 1108–1114*

[18]M. I. Mendelson; Average grain size in polycrystalline ceramics, *J. Am.Ceram. Soc., 52 (1969) [8] 443–6*

COMBUSTION SYNTHESIS (SHS) OF COMPLEX CERAMIC MATERIALS

Jerzy Lis
AGH University of Science and Technology
Faculty of Materials Science and Ceramics
Al. Mickiewicza 30, 32-050 Kraków, Poland

ABSTRACT

The present work is focused on efficient and convenient powder processing by using combustion synthesis also called self-propagating high-temperature synthesis (SHS). A background for materials preparation using SHS is discussed. Next a review of different forms of complex ceramic materials obtained directly by SHS or using SHS-origin precursors is presented. The use of SHS may bring a considerable impact in ceramic technology, by enabling a manufacturing of high-purity and phase-controlled powders. Rapid combustion conditions were successfully used to manufacturing complex ceramic powders very active in sintering and suitable for preparing of multiphase complex materials. It can be demonstrated in different ceramic systems explored by the author and co-workers using SHS e.g. Si-C-N, Ti-Al-C-N, as well as Al-O-N. It has been concluded, that the SHS technique has brought a contribution to the ceramic processing and still may be considered as a perspective approach for materials engineering.

COMPLEX MULTIPHASE MATERIALS PREPARED BY COMBUSTION SYNTHESIS (SHS).

Unconventional challenges in materials preparation is accompanied with combustion synthesis also called self-propagation high-temperature synthesis (SHS). In order to proceed with the SHS, an exothermic reaction and self-sustaining character of the process are required. The reaction, locally ignited in the raw mixture, results in a formation of a combustion wave, which propagates through the reaction system transforming it into final product. Utilization of intrinsic heat of the reaction, as well as, short time and relatively simple equipment make SHS as an economic and competition to others methods for production of a great number of materials. Starting from the first pioneer works of Merzhanov and co-workers [1,2] SHS has been successfully employed for many materials technologies in laboratory, as well as, in industrial scale. Because of non-limited possibilities in creating of SHS phenomena, hundreds of different materials are reported as prepared using SHS, in thousands papers published during almost 50 years of SHS history.

Development of SHS techniques results in two basic methods of preparation of complex multiphase materials. In the first, the materials are created „in situ" during SHS combustion. The second treats SHS products as a transient material i.e. precursor which is processed using following technological operations to the final materials form. "In situ" preparation is the natural form of SHS technique. A common form of SHS- prepared complex multiphase products is a polycrystalline structure with a grain size of few micrometers. Because of homogenous form of starting powder mixing, and stochastic character of combustion physicochemical phenomena the dispersed phases are usually uniformly distributed in product volume creating a materials which can be called "particulate composite". Significant progress observed in SHS science results in better understanding of SHS phenomena. It leads to improved of control of products microstructure and properties by controlling, first at all, of burning velocity, combustion temperature and extent of conversion. The combustion phenomena can be also regulated by green materials parameters (composition, particle size, density, charge volume, initial temperature, type and amount of additives and fillers, etc.) as well as combustion conditions (composition and pressure of ambient gases, external influences).

SHS-derived precursors are prepared mainly in form of powders next processed into polycrystals using different densification techniques. The sinterable powders can be commonly obtained by deep grinding and milling of SHS products. The alternative approach is focused on preparation of fine powders with controlled morphology direct by SHS combustion [3]. Then, the powders are densified with or without sintering additives by presureless sintering; hot pressing (HP) or hot-isostatic pressing (HIP) [4]; spark plasma sintering (SPS) [5] or thermal spraying coatings (TS) [6]. The powders can be additionally mixed with other components like, powders, whiskers, fibres, etc. for obtain more complex composite materials also in form of functional gradient materials (FGM) [7]. In many cases, SHS derived powders are found as the more sinterable compare to others prepared using more conventional techniques [8,9]. It can be explain because of their more reactive state related with non-equilibrium phases or defected structures created during high-temperature rapid SHS phenomena. Then, the sintering is accompanied by additional chemical reactions and/or phase transformation in complex reactive sintering [10].

COMPLEX CERAMICS PROCESSED FROM SHS-DERIVED POWDERS

SHS Research in the AGH UST

The first works on SHS field in AGH-UST, Kraków were done by Pampuch and co-workers in '70 last century, when they investigated combustion phenomena during processing of SiC-Si composites prepared by high-temperature filtration in Si-carbon fibres system [11]. Investigations in SHS were continued and majority of the works were focused on preparation of sinterable ceramic powders of hard and refractory compounds using different combustion techniques.

SHS is especially useful for preparation of non-oxide ceramic powders like carbides, nitrides, borides, etc. i.e. compounds having generally rather low sinterability. The investigations showed, that a proper utilization of SHS in preparation of these powders, can lead to sinterable ones. An analysis of the experimental examples indicates that sintering phenomena of such a way prepared powders have mostly a character of densification promoted by reaction in solid and/or liquid phase. It can result from specific combustion-type phenomena during SHS, where the powders can be prepared in non-equilibrium forms including: metallic dopants, transient phases having defectable structure, etc. This can improve their sinterability during following high-temperature treatment by forming the transient liquid or s.s. phases. The basic problem is understanding and controlling of these phenomena.

The aim of the AGH-UST works was preparation of single- phase sintered bodies as well as multiphase complex materials starting from SHS powders. In the last area, the most interesting results in form of complex materials with interesting properties have been achieved for materials in several explored systems: Si-C-B [12]; Si-C-N [13]; Si-Al-O-N [14,15], Si-Al-N-C [16]; Al-O-N [17], Ti-B-C [18], Ti-Si-C [19,20], Ti-Si [21]; Ti-Al-C [22]; Ti-Si-C-N [23] and Ti-Al-C-N [24]. In the present paper the review of results collected in three systems Si-C-N; Ti-Al-C-N and Al-O-N are presented.

Materials Prepared in the Si-C-N System.

The studies on the silicon carbide formation by solid combustion in the Si-C-N system were carried out to prepare the composite $SiC-Si_3N_4$ and pure SiC nano-powder [25-27]. Conventional SHS combustion in the Si-C system provides products in form of strongly agglomerated SiC coarse grains demanding long time of milling for further processing. In contrast realization of the

Si+C exothermic reaction under nitrogen pressure gives much more efficient way of silicon carbide nanopowder preparation.

Combustion reaction propagates spontaneously in self sustaining regime for all nitrogen pressures. The peak temperature of combustion is presented in the Figure 1.

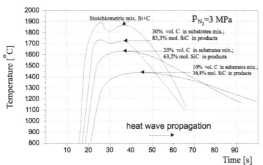

Fig. 1. Peak temperature during combustion in the Si-C-N system

The combustion products for 2.5 MPa; and 4.0 MPa of N_2 are consisted of β-SiC and β-Si_3N_4 phases. In case of 3.0 MPa and 3.5 MPa of nitrogen pressure pure β-SiC phase is detected only. The absence of free silicon peaks for all nitrogen pressures indicated a high reaction degree, estimated to be higher than 98%.

The specific surface area of products with associated the maximum peak temperature of combustion in the Si-C-N system are shown in Fig.2. A gradual increase of specific surface area with increasing nitrogen pressure up to 3,5 MPa is observed. However, higher pressure of 4,0 MPa resulted in decreasing of products dispersion. It is interesting to note that specific surface area of powders well corresponded to phase composition of products. The final pure SiC powder have the highest specific surface area of about $19m^2/g$ but products with silicon nitride content obtained under 2.5 MPa and 4.0 MPa of nitrogen pressure showed lower specific surface area of about 8-12 m^2/g. Assuming that the measured surface area is the sum of the areas of silicon nitride and silicon carbide powders, we see that synthesized SiC is much finer than Si_3N_4. The reaction maximum temperature is gradually increasing as nitrogen pressure is also increasing but high temperature of reaction does not influence the surface area of powdered products as it is commonly observed for the combustion synthesis of pure β-Si_3N_4.

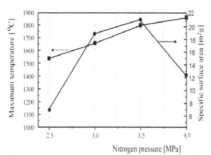

Figure 2. Specific surface area and maximum temperatures vs. nitrogen pressure of SHS combustion in the Si-C-N system

In Figure 3a and 3b the morphology products obtained under 3.0MPa and 4,0MPa are shown respectively. Based on EDS analysis and XRD measurements, it is found that nanometric particles in Fig. 3a are silicon carbide phase. Thus, we concluded that SHS reaction accomplished under 3.0MPa of nitrogen pressure resulted in silicon carbide nanopowder. A bi-modal size distribution of particles can be observed in Fig.2b. The larger particles are faceted and have sharp edges, while the smaller ones are still in nanometric size in the form of agglomerations. The presence of silicon nitride phase leads to conclusion that coarser grains are β-silicon nitride.

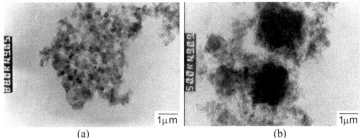

(a)   (b)

Figure 3 (a) TEM images of silicon carbide nano-powder synthesized under 3.0 MPa of nitrogen pressure; (b) silicon carbide - silicon nitride nanopowder synthesized under 4.0 MPa of nitrogen pressure.

The results of this study, and investigation of Kata et al into combustion synthesis in the Si-C-N system indicated the complex character of the combustion which is controlled by silicon nitride decomposition [28].

In the Si-C-N system temperature curves for 2,5 MPa and 4,0 MPa of nitrogen pressure have one-modal flat character with maximum temperatures reaching 1530°C and 1880°C respectively. For 3.0 MPa and 3.5 MPa of nitrogen pressure temperature profiles assumes a two-modal character with maximum temperatures reached 1650°C and 1800°C respectively. Phase composition of combustion products is well correlated with combustion temperature profile. If temperature curves are bi-modally shaped the reaction products consist of pure silicon carbide. In contrast combustion having one modally shaped temperature resulted in silicon nitride remains in products.

Obtained powders can be sintered to dense materials from single phase SiC to complex Si_3N_4-SiC particulate composites as we can see in Fig. 4.

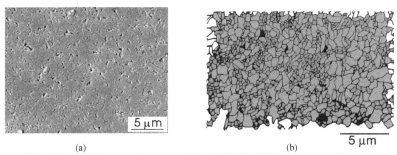

(a)   (b)

Fig. 4. (a) Si_3N_4-25%SiC complex material; (b) digitalized phase composition - more dark are marked SiC grains.

Complex Materials with MAX Phases in Ti-Si-C and Ti-Al-C-N Systems

Among many covalent materials such as carbides or nitrides there is a group of ternary compounds referred in literature as MAX-phase, H-phases, Hägg-phases, Novotny-phases or nanolaminates. These compounds have a $M_{n+1}AX_n$ stoichiometry, where M is early transition metal, A is an element of A groups (mostly IIIA or IVA) and X is no-metal, mainly carbon and/or nitrogen. Heterodesmic structures of these phases are hexagonal and specifically layered.

The specific structure of such compounds causes that they show an extraordinary set of properties that are not observed in any typical metallic and ceramic material [29]. The MAX materials combine common ceramic properties like high stiffness, moderately low coefficient of thermal expansion and excellent thermal and chemical resistance with low hardness, good compressive strength, high fracture toughness, ductile behaviour as well as good electrical and thermal conductivity characteristic for metals. The exceptional combination of such properties allows MAX materials to be good machinable even by standard tools.

The most successful preparation method of MAX powders has been developed by the authors based on self-propagating high-temperature synthesis (SHS). The first explored system was Si-Ti-C, as it is shown in Fig. 5 [19]. Using powders of titanium, silicon and carbon black as reactants a powder composed mainly of Ti_3SiC_2 and of minor amounts of TiC was synthesized. Following ignition at about 1050°C, complete conversion of the reactants to the product was observed in a time of 2 to 5 s. [30].

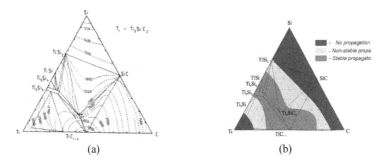

(a) (b)

Fig. 5. Combustion in the Si-Ti-C system: (a) theoretical combustion temperatures, (b) experimental combustion phenomena

The SHS-derived powders are good precursors for preparation of dense materials. A typical powders synthesized through SHS have a form of a porous sponge. Because of strong aggregation and agglomeration of the powders caused mainly by high temperature of SHS reaction grinding process is usually applied before densification. Pressureless sintering, hot-pressing and isostatic hot-pressing were successfully applied to obtain dense materials with MAX as a major phase [31,32]. The densification temperatures differ from 1100 to 1600°C depending on the process and the powder. It is worth to notice that during thermal treatment some structural changes accompanied with phase content changes can be observed. Controlling of starting SHS-derived powders composition, as well as sintering conditions can bring to obtaining of different materials from single-phase Ti_3SiC_2 polycrystals to Ti_3SiC_2-TiC complex composites [33] (see Fig. 6).

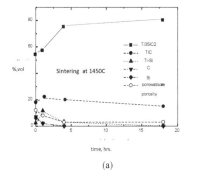
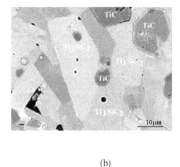

(a)   (b)

Fig. 6. (a) Phase composition during sintering of SHS obtained Ti-Si-C powders (b) final Ti₃SiC₂-TiC complex materials

The most interesting features of obtained MAX-based materials are their unique mechanical properties. High fracture toughness and possibility of plastic deformation is result of complex fracture mechanism additionally depending on microstructure of the material, especially on grain sizes and their shapes. They had been successfully tested as armour materials also in the FGM system, see Fig. 7 [34].

Fig. 7. FGM Al₂O₃-Ti₃SiC₂ armour ceramics elements

More perspective because of specific properties are new MAX-based complex materials explored in the systems: Ti-Si-C-N and Ti-Al-C-N. For instance, heat-treatment of SHS derived powders from Ti-Al-C-N system lead to formation of complex materials contained MAX phases combined with TiC and TiN [35-37]. Different type of composite materials were prepared, namely Ti₂AlC-TiC, Ti₃AlC₂-TiC, Ti₂AlN-Ti₃AlN-TiN and Ti(C, N)-Al. Multi stage SHS procedure was applied for manufacturing of powders. First stage was synthesising intermetallic precursor in Ti-Al system by thermal explosion, then this precursor was used for local ignition initiated SHS of more complex materials.

Mixed powders were homogenized and after that process they were placed in a high-pressure reactor as a loose bed in a graphite holder. The synthesis of Ti₃AlC₂-TiC and Ti₂AlC-TiC were performed in argon atmosphere, while in the case of Ti₂AlN-TiN and Ti(C, N)-Al was performed at 0.5 MPa of nitrogen. The SHS synthesis was initiated by a local ignition. Process was very rapid and last for few minutes (depends on volume of reactants). Obtained products had character of dense, sintered materials. Powders were crushed and ground in an isopropanol medium in rotary-vibratory mills and hot-pressed in temperature range of 1300-1850°C in a constant argon or nitrogen flow. Dense materials, characterized by porosity below 1% were obtained. In case of each of materials chemical reactions were undergoing during the reaction sintering process. Obtained materials are characterised by interesting mechanical properties with low hardness and high structure toughness. Their characteristic layered structure is presented on Fig. 8 and 9.

Fig. 8. Composite Ti_3AlC_2-TiC

Fig. 9. Composite Ti_2AlN-Ti_3Al-TiN

Materials in the Al-O-N System.

Different complex multiphase materials in the Al_2O_3-AlON system were prepared using two steps procedure with SHS used as the method for preparation of powdered precursors. The filtration combustion technique was explored by burning of aluminium metal and corundum powders mixtures in nitrogen atmosphere, according to the general reaction: $2Al + Al_2O_3 + N_2 = Al_3O_3N + AlN$.

Details of experimental procedure was described elsewhere [38,39]. The mixture of reactants was introduced into the chamber in the form of homogeneous, loose powder bed poured into graphite crucible. The combustion was initiated by electrical pre-heating of the graphite crucible by electric current flow for several seconds. The combustion got start rapidly, the reaction mixture reached temperature exceeded 2000°C in few seconds and after next about 60 second the reaction was completed. Optimum conditions were found by experimental procedure let to obtain >99% yield of reaction. The phase composition of the combustion products and their morphology were determined, as they are presented in Fig. 11 and Fig. 12, respectively.

The SHS synthesized powders were composed of aluminium nitride, γ-alon and not completely reacted corundum with different proportions. The materials had sponge character with non-uniform aggregated and agglomerated morphology having some elements typical for gas-phase reaction. The SHS products were crushed and next milled in a rotary-vibratory mill. The sintering experiments were conducted in graphite resistance furnace in flowing nitrogen at different temperatures with temperature and soaking at maximum temperature for 2 h.

The results showed that the obtained powders were sinterable and maximum densities >98% were achieved. Uniform microstructure with an average grain size about 3.5 μm is observed.

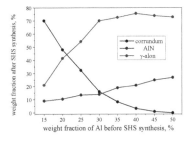

Fig. 10. Composition of SHS derived powders   Fig. 11. SHS product morphology

The detail studies of phase composition during sintering showed that densification processes were coupled with several chemical reactions changing powder phase composition (see. Fig. 12) [40].

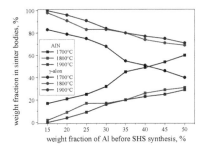

Fig. 12. Changes of phase composition of AlN - γ-alon materials sintered at different temperatures.

Finally, only two-phase materials composed of AlN and γ-alon are observed with different ratios controlled by composition of the starting powder mixture and sintering temperature. Complex materials in the AlN-γalon system prepared using SHS-derived powders have advantaged mechanical, thermal, electrical properties [41] as well as high corrosion resistance against molten metals, especially nickel-based super-alloys [42].

CONCLUSIONS

It can been concluded, that the SHS technique has brought an important contribution to the materials processing and may be considered as perspective approach for preparation of complex multiphase ceramic materials.

ACKNOWLEDGMENT
The author wants to thank the co-workers: M. Bucko; L. Chlubny; D. Kata and D. Zientara for important impact in preparation of this paper as well as the National Science Centre, Poland for the financial support in the project no. 2472/B/T02/2011/40.

REFERENCES
[1] Merzhanov A.G., Borovinskaya I.P. *Self-propagating high-temperature synthesis of inorganic compounds,* Dokl. Akad. Nauk SSSR. 1972, vol. 204, 2, 366-369

[2] Merzhanov A.G., *Self-propagating high-temperature synthesis: Twenty years of search and finding,* in Combustion and Plasma Synthesis of High-temperature materials, Eds. Munir Z.A., Holt J.B. at all, VCH Publ., 1990, 1-53

[3] Amosov A.P., Borovinskaya I.P., Merzhanov A.G., Sytschev A.E. *Pronciples and methods for Regulation of Dyspersed Structure of SHS Powders: From Monocrystallites to Nanoparticles,* Int. J. of SHS, 14 (3) 2005 , 165-185 [5] Zhang L., Zhao Z.M., Liu W.Y., Lu H.X. *High-Gravity Activated SHS of Large Bulk Al$_2$O$_3$/ZrO$_2$ Nanocrystallline Composites,* Int.J. of SHS, 18, 2009, 173-1280

[4] Capel F., Contreras L, Rodriges M.A., *Mechanical Behaviour of hard Ceramic Based Composites,* Key Engineering Materials Vols. 264-268, 2004, 1025-1028

[5] Tsuchida T., Yamamoto S., *Spark Plasma Sintering of ZrB$_2$-ZrC Powder mixtures Synthesized by MA-SHS in Air,* J. Mat. Sci. 42 (3), 2007, 772-778

[6] Talako T., Ilyuschenko A., Letsko A., *SHS Powders for Thermal Spray Coatings,* KONA Powders and Particle Journal, 27, 2009, 55-72

[7] Łopaciński M., Lis J., *Ceramic Functionally Gradient Materials for armour applications,* ECerS Proc. of the 10th Int. Conference of the European Ceramic Society : June 17–21, 2007, eds. J. G. Heinrich, C. G. Aneziris. — Baden-Baden : Göller Verlag GmbH, 2007, 1279–1284

[8] Ermer E., Lis J. , Pampuch R., *Investigation of Sialon Powders and Sintered Materials by FTIR Method,* Proc. Fourth Euroceramics, C. Gallasi (ed.), C.N.R.-IRTEC, Faenza Italy, Vol. 1, 1995, 61-66

[9] Pampuch R. , Lis J., *Sinterable SHS Powders. Illustrative Examples of State-of-the-Art,* Adv. Sci. and Tech.; 45 (2006) 969-978

[10] Lis J. *Sinterable powders of covalent compounds prepared by SHS,* Ceramics 44[4], 1994, 1-74

[11] Pamuch R., Białoskórski J., Walasek E. *Mechanism of reactions in the Si$_l$ + C$_f$ system and the self-propagating high-temperature synthesis of silicon carbide* Ceram Int. 13 (1) 1987 63-67

[12] L. Stobierski, E. Ermer, R. Pampuch and J. Lis, *Supersaturated Solid Solutions of Boron in SiC by SHS* Ceramics Int. 19 (1993) 231-234

[13] R. Pampuch, L. Stobierski, J. Lis   Mictrostructure Development   on Sintering of SHS-Derived and Conventional Silicon Carbide and Nitride Powders" Int. J. of SHS 2[3] (1993) 159-164

[14] J.Lis, S. Majorowski, J.A. Puszynski, V. Hlavacek, *Dense β and α/β-Sialon Materials by Pressureless Sintering of Combustion Synthesized Powders,* Ceram. Bull., 70, [10], (1991), 1658-1664

[15] Kata D., Lis J., Pampuch R. *Nitrogen Powders Prepared by Combustion Methods,* Ceramics Ceramika 45, 1994, 28-35

[16] Stobierski L., Lis J., Węgrzyn Z., M. Bućko M.M., *SHS Synthesis of Nanocomposite AlN-SiC Powders,* Int. J. of SHS, 10 (2001) 217-226

[17] Zientara D, Bućko M.M.; Lis J.; *Dense gamma-Alon Materiale from SHS Synthesized Powders*; Adv. Sci. and Tech.; 45 (2006) 1052-1057

[18] R. Pampuch, J. Lis and L. Stobierski, "Solid Combustion Synthesis of Silicon-Containing Materials in the Presence of Liquid Silicon Alloys", Intern. Journal of SHS, 1, [1], (1992) 78-82

[19] Pampuch R., Lis J., Stobierski L., M. Tymkiewicz M., Solid Combustion Synthesis of Ti_3SiC_2, J. Europ. Ceram. Soc. 5 (1989) 283-287

[20] Lis J., Miyamoto Y., Pampuch R., Tanihata K., Ti_3SiC_2 -based Materials Prepared by HIP-SHS Techniques, Materials Letters 22 (1995) 163-168

[21] Szwagierczak D., Marek A, Gadurska J., Kulawik J., Lis J. Use of Various Titanium Silicides to Thick Film Resistive Pastes" Proc. XXIIIIMAPS, Kołobrzeg 21-23 Sept. 1999, 149-154

[22] M. Łopaciński M., Puszyński J., Lis J., Synthesis of Ternary Titanium Aluminum Carbides Using Self-Propagating High-Temperature Synthesis Technique, J. Amer. Ceram. Soc. 84 [12] (2001) 3051-3053

[23] Lis J., Kata D., Chlubny L., Łopaciński M., Zientara D., Processing of titanium-based layered ceramics using SHS technique, Ann. Chim. Sci. Mat. 2003, 28 (Suppl. 1), S115-S122.

[24] Chlubny L., Lis J., Bućko M.M.; SHS Synthesis of the Materials in the Ti-Al-C-N System Using Intermetalics; Adv. Sci. and Tech.; 45 (2006) 1047-1051

[25] Kata D., Lis J., Pampuch R., Stobierski L., Ermer E., "Preparation of Si_3N_4-SiC composite powders by combustion in the Si-C-N system" Arch. Combustionis 16, 1-2 (1996) 13-21.

[26] Kata D., Lis J., Pampuch R. and Stobierski L. "Preparation of Fine Powders in the Si-C-N system using SHS" Int. J. of SHS, 7, 4, (1998) 475-485.

[27] Kata D., Lis J., Pampuch R., "Combustion Synthesis of Multiphase Powders in the Si-C-N System" Solid State Ionic 101-103, (1997) 65-70.

[28] Kata D., Lis J., "Ceramic Composites in the Si_3N_4-SiC System" Archives of Metallurgy, vol. 42, no. 2 (1997) 33-141.

[29] Jerzy Lis, Leszek Chlubny, Michał Łopacinski, Ludosław Stobierski, Mirosław M. Bućko Ceramic nanolaminates – processing and applications, Journal of the European Ceramic Society, 2008 vol. 28 s. 1009–1014

[30] J. Lis, R. Pampuch and L. Stobierski," Reactions during SHS in the Ti-Si-C System" , Int. J. of SHS 1 [3] (1992) 401-408

[31] J. Lis, Y. Miyamoto, R. Pampuch, K. Tanihata "Ti3SiC2 -Based Materials Prepared by HIP-SHS Techniques" Materials Letters 22 (1995) 163-168

[32] R. Pampuch, J. Lis "Ti_3SiC_2 - A Plastic Ceramic Material", Adv. In Science and Tech. 3B, P. Vincenzini (ed.), Techna Srl, Faenza 1995, 725-732

[33] M. Faryna, J. Lis, R. Kórnik "SEM studies of microstructural development during sintering of Ti3SiC2-based composites" J. Trace and Microprobe Techniques 15(4) (1997) 453-457

[34] M. Łopaciński, J. Lis „New Funcjonally Gradient Materials in the Ti-Si-C System" Powder Metallurgy 3-4 (1999) 42-45

[35] Chlubny L., Lis J., Bućko M.M. Preparation of Ti_3AlC_2 and Ti_2AlC powders by SHS method, Materials Science & Technology 2009, October 25-29, 2009 : Pittsburgh, Pennsylvania. S., The Printing House, Inc., 2205—2213

[36] Chlubny L., Lis J., Bućko M.M. Phase evolution and properties of Ti_2AlN based materials, obtained by SHS method" – Proc. of the 32nd International Conference on Advanced Ceramics and Composites, Wiley, 2008, pp 13-21

[37] Lis J., Chlubny L., Zientara D., Bućko M.M. Phase evolution of materials in the Ti-Al-C system during hot pressing, Inżynieria Materiałowa 3-4 (156-157), 2007

[38] D. Zientara D, Bućko M.M., Lis J.. Dense γ-alon materials derived from SHS synthesized powders. Adv. Sci. Tech. 45 (2006) 1052-1057.

[39] Zientara D., Bućko .M.M, Lis J.. Alon-based materials prepared by SHS technique. J. Eur. Ceram. Soc. 27 (2007) 775-779.

[40] Zientara D., Bućko M.M., Lis J.. *Investigation of γ-alon structural evolution during sintering and hot-pressing*. Key Eng. Mater. 409 (2009) 313-316.

[41] Zientara D., Bućko M.M., Lis J.. *Dielectric properties of aluminium nitride – γ-alon materials*. J. Eur. Ceram. Soc. 27 (2007) 4051-4054

[42] Zientara D, Bućko M.M., Lis J. *Aluminium oxynitride as a crucible material for melting of nickel-based superalloys*. in Proceedings of the 10th International Conference of the European Ceramic Society, edited J.G. Heinrich, C.G. Aneziris, Göller Verlag, Baden-Baden (2007), pp. 1882-1885.

WEAR AND REACTIVITY STUDIES OF MELT INFILTRATED CERAMIC MATRIX COMPOSITE

Jarmon, D. C., and Ojard, G.C.

United Technologies Research Center, East Hartford, CT

ABSTRACT

As interest grows in the use of ceramic matrix composites (CMCs) for critical gas turbine engine components, the effects of the CMCs interaction with the adjoining structure needs to be understood. A series of CMC/material couples were wear tested in a custom elevated temperature test rig and tested as diffusion couples, to identify interactions. Specifically, melt infiltrated silicon carbide/silicon carbide (MI SiC/SiC) CMC was tested in combination with a nickel-based super alloy, Waspaloy, a thermal barrier coating, yttria stabilized zirconia (YSZ), and a monolithic ceramic, silicon nitride (Si_3N_4). To make the tests more representative of actual hardware, the surface of the CMC was kept in the as-received state (not machined) with the full surface features/roughness present. Test results include: scanning electron microscope characterization of the surfaces, micro-structural characterization, and microprobe analysis.

INTRODUCTION

Ceramic matrix composites (CMCs) have decades of development and testing efforts behind them [1-4]. With this background and key material attributes, CMCs are also finding technical applications [5,6]. As technical acceptance is achieved, additional challenges and insertion issues are discovered resulting in additional efforts and insight. One area that is coming to light as testing proceeds is the material interaction with the supporting structure and contacting hardware.

As part of the initial effort to understand such interactions, a melt infiltrated silicon carbide/silicon carbide (MI SiC/SiC) CMC was used in a series of wear and reactivity testing. The wear testing was done at temperature against a series of materials that could be envisaged as possible materials that would be present in actual applications. These materials were Waspaloy, yttria stabilized zirconia (YSZ) and silicon nitride. In addition, material couples were made to note material interactions. The effort focused on characterization after select times to see if micro-structural changes were present (mainly surface related).

PROCEDURE

Material

As noted, the material for this study was the MI SiC/SiC CMC system. This material has been studied by other investigators [7-9]. The MI SiC/SiC composite is comprised of a stoichiometric SiC fiber (Sylramic™) with a boron nitride (BN) interface coating. The SiC matrix is infiltrated by vapor deposition followed by slurry casting of SiC particulates and a final melt infiltration of Si. The fiber preform was a cross-ply balanced 5 harness satin weave with 18 ends per inch and 36% fiber volume fraction. The material cross section (micro-structure) has been shown by other investigators [8,9].

Wear Testing

Wear testing was done using a wear rig developed by United Technologies Research Center (UTRC). A schematic of the wear rig is shown in Figure 1. This unit consists of a fixture that holds two materials in contact inside a furnace capable of 1200°C. The fixture permits loads up to 36 N to be applied to the specimens while simultaneously imposing displacements at frequencies ranging from 0.005m at 100 Hz to 0.0127m at 20 Hz. The tests were performed under a flowing oxygen atmosphere at various temperatures. The experimental series is shown in Table I. For this testing series, the surface of the MI SiC/SiC was left in the as-processed state with the surface finish not machined to remove any surface asperities. All samples were 25 mm x 25 mm for the faces in contact.

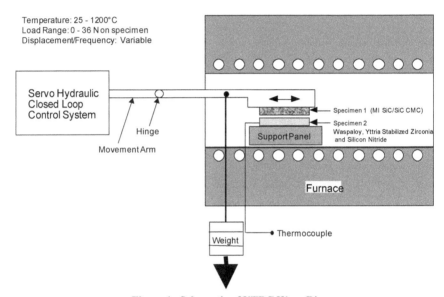

Figure 1. Schematic of UTRC Wear Rig

Table I. Wear Testing Experiment Series

Specimen 1	Specimen 2	Temp (°C)	Time (hr)	Displacement (mm)	Hz	Pressure (MPa)
MI SiC/SiC	Si_3N_4	1038	100	0.1	90	0.055
MI SiC/SiC	YSZ	1010	100	0.1	90	0.055
MI SiC/SiC	Waspaloy	649	300	0.1	103	0.055

Reactivity Testing

Reactivity tests were run in an alumina tube furnace with flowing oxygen atmosphere (2.0 · 10^{-5} m³/s or 1.2 liter/min). The experimental test plan is shown in Table II. Close contact between the samples for reactivity testing was done by having a weight on top of the sample stack. This weight was achieved using an alumina plate and resulted in an estimated average pressure of 1.8 MPa. Samples were removed from the furnace after the furnace had cooled to room temperature. After furnace exposure, select samples were sectioned for micro-structural characterization. All samples were 25 mm x 25 mm for the faces in contact.

Table II. Reactivity Testing Experiment Series

Specimen 1	Specimen 2	Temp	Time
		(°C)	(hr)
MI SiC/SiC	Si_3N_4	1038	500
MI SiC/SiC	Si_3N_4	1038	1000
MI SiC/SiC	Si_3N_4	1038	4000
MI SiC/SiC	YSZ	1038	1000
MI SiC/SiC	YSZ	1038	2000
MI SiC/SiC	YSZ	1038	4000
MI SiC/SiC	Waspaloy	649	1500
MI SiC/SiC	Waspaloy	649	4000

Characterization Post Exposure:

Selected samples from both the wear and reactivity studies were laser cut and a cross section was mounted and polished for optical microscopy, scanning electron microscopy (SEM) and microprobe analysis. In some cases, the contacted surface of the sample was also examined.

RESULTS

Wear

No visible amount of wear was seen on the surface of the MI SiC/SiC against Si_3N_4 for both materials. There was a white coating observed on the surface of the samples. SEM and microprobe analysis is shown in Figure 2. The surface coating was most likely silica, SiO_2.

The wear couple between the MI SiC/SiC and YSZ showed an area of wear. The wear was present in an area where the asperity peaks of the MI SiC/SiC rubbed on the YSZ. This is shown in Figure 3. There was little material transfer of the YSZ to the MI SiC/SiC. Detailed microprobe analysis showed that a small portion of the YSZ was scrapped off and that is consistent with the images shown in Figure 3. In addition, oxidation of the surface of the MI SiC/SiC was observed.

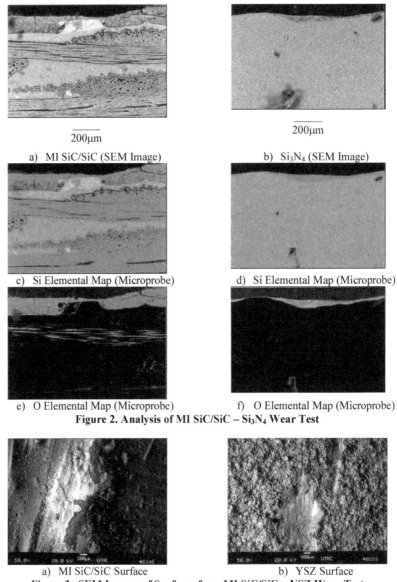

200μm

a) MI SiC/SiC (SEM Image)

200μm

b) Si₃N₄ (SEM Image)

c) Si Elemental Map (Microprobe)

d) Si Elemental Map (Microprobe)

e) O Elemental Map (Microprobe)

f) O Elemental Map (Microprobe)

Figure 2. Analysis of MI SiC/SiC – Si₃N₄ Wear Test

a) MI SiC/SiC Surface

b) YSZ Surface

Figure 3. SEM Images of Surfaces from MI SiC/SiC – YSZ Wear Test

The wear between the MI SiC/SiC and Waspaloy showed more areas of wear (8 areas on the 25 mm x 25 mm surface) where the MI SiC/SiC peak surface roughness came into contact of the surface of the Waspaloy. The localized pressure was determined to be on the order of 27 to 69 MPa based on the applied load and size of the surface asperities. The areas of wear were found to be 12.7 μm deep (via surface profilometry). The corresponding areas on the MI SiC/SiC showed some discoloration but no wear. It was clear that material was transferred between the wear couples. Figure 4 shows the microprobe analysis of the surface of the MI SiC/SiC. There was a metal oxide layer present with the atomic species consistent with the Waspaloy material sample. In addition, there were suspected silicide layers present on the Waspaloy, as shown in Figure 5.

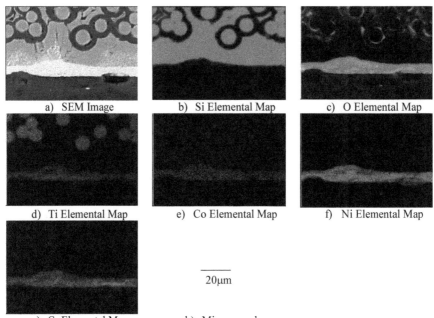

a) SEM Image   b) Si Elemental Map   c) O Elemental Map

d) Ti Elemental Map   e) Co Elemental Map   f) Ni Elemental Map

20μm

g) Cr Elemental Map   h) Micron marker

Figure 4. Analysis of MI SiC/SiC Surface After Wear Test with Waspaloy

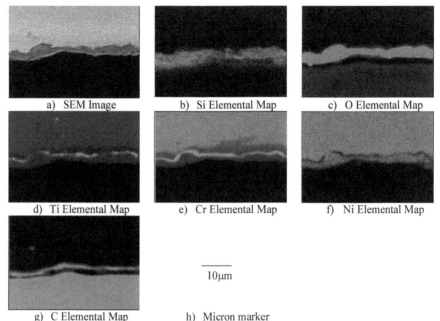

a) SEM Image	b) Si Elemental Map	c) O Elemental Map
d) Ti Elemental Map	e) Cr Elemental Map	f) Ni Elemental Map
g) C Elemental Map	h) Micron marker	10μm

Figure 5. Analysis of Waspaloy from Wear Test with MI SiC/SiC

Reactivity

There was no evidence of any reactivity of the MI SiC/SiC against Si_3N_4 and YSZ and the only thing to note was the presence of the silica scale on the MI SiC/SiC as was seen in the wear efforts. This is consistent with the relatively low temperature present and the preponderance of ceramic species in the effort.

More complex oxidation product was seen on the reactivity couple of the MI SiC/SiC against the Waspaloy. The Waspaloy generated a Cr oxide scale which is consistent with expectations for such a metal system in an oxidizing environment. [10]. In addition, there were areas in the oxide scale that were rich in silicon as shown in Figure 6. The silicon was measured to be about 2 μm in thickness. This may be due to areas where the surface roughness of the MI SiC/SiC allowed point contact. As a result, silicon was diffused from the SiC/SiC into the Waspaloy. Only a thin film of silica was formed on the surface of MI SiC/SiC, even after 4000 hrs of exposure which is consistent with long term exposure (see Table II).

The nature of the point contact can be seen by looking at the surface of the MI SiC/SiC after 4000 hrs as shown in Figure 7, looking at just one of the species found in the Waspaloy. Figure 7 shows the element Cr and the surface map shows the contact nature of the test and the effect of the surface roughness as well as the mobility within the Si phase.

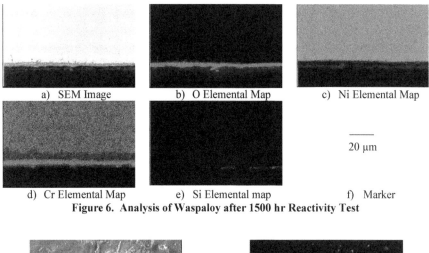

a) SEM Image b) O Elemental Map c) Ni Elemental Map

20 μm

d) Cr Elemental Map e) Si Elemental map f) Marker

Figure 6. Analysis of Waspaloy after 1500 hr Reactivity Test

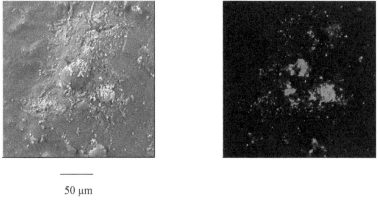

50 μm

a) SEM b) Cr Elemental Map

Figure 7. Analysis of MI SiC/SiC Surface after 4000 hr Reactivity Test

DISCUSSION

The wear and reactivity testing of ceramic species against ceramic species only showed limited concern due to the surface condition of the as-processed MI SiC/SiC surface (Figure 3). There was some formation of oxide (silica) scale but that was not unexpected (and predicted based on thermodynamic calculations) [11] (Figure 2). When considering the presence of a metal present in the couple, other researchers of metal/ceramic couples have seen that Ni is the major species that diffuses into the Si based carbide ceramic [12]. They have also shown that Si and C move into the metal. The presence of Si diffusing in was seen in this effort (Figure 6).

The difference in this study is the presence of Si as part of the melt infiltrated process that is a metal in the structure. It is not carbide. This provides a path for species to move within the composite as shown in previous work of one of the authors [9]. This is seen in the Cr elemental map of Figure 7, where Cr is present on the surface and has diffused. It was surmised that the Si provides a path for diffusion.

CONCLUSION

No significant reaction occurred at the interface for the various material couples under the current reactivity tests except for some oxidation and minor reaction (SiC/SiC-Waspaloy). The wear tests have shown that small amount of wear and reaction occurred under the current experimental conditions. In most cases, thermodynamic calculation as well as previous experimental studies could account for the present experimental observations. There is concern about the Si present in the material from the melt infiltrated process acting as a transport media to facilitate the migration of chemical species within the MI SiC/SiC.. This later point warrants additional investigation with polished faces present to assure greater contact between the samples to clearly note if Si provides a preferred path for diffusion.

ACKNOWLEDGMENTS

The authors are indebted to Dr. Hong Du, formerly of United Technologies Research Center, for her assistance in performing and organizing the work presented in this paper. Her help is greatly appreciated. This work was performed under the Enabling Propulsion Materials Program, Contract NAS3-26385, Task A, David Brewer program manager.

REFERENCES

1. Richerson, D.W., Iseki, T., Soudarev, A.V. and van Roode, M., "An Overview of Ceramic Materials development and Other Supporting Technologies", Chapter 1 from Ceramic Gas Turbine Component Development and Characterization, van Roode, M., Ferber. M.K. and Richerson, D.W., Eds. ASME Press, New York, 2003
2. Landini, D.J., Fareed, A.S., Wang, H., Craig, P.A. and Hemstad, S., "Ceramic Matrix Composites Development at GE Power Systems Composites, LLC", Chapter 14 from Ceramic Gas Turbine Component Development and Characterization, van Roode, M., Ferber. M.K. and Richerson, D.W., Eds. ASME Press, New York, 2003
3. Brewer, D., 1999, "HSR/EPM Combustion Materials Development Program", Materials Science & Engineering, 261(1-2), pp. 284-291.
4. Brewer, D., Ojard, G. and Gibler, M., "Ceramic Matrix Composite Combustor Liner Rig Test:, ASME Turbo Expo 2000, Munich, Germany, May 8-11, 2000, ASME Paper 2000-GT-670.
5. Kestler, R. and Purdy, M., "SiC/C for Aircraft Exhaust", presented at ASM International's 14th Advanced Aerospace Materials and Processes Conference, Dayton, OH, 2003
6. K.K. Chawla (1998), *Composite Materials: Science and Engineering, 2nd Ed.*, Springer, New York.

7. J.A. DiCarlo, H-M. Yun, G.N. Morscher, and R.T. Bhatt, "SiC/SiC Composites for 1200°C and Above" Handbook of Ceramic Composites, Chapter 4; pp. 77-98 (Kluwer Academic; NY, NY: 2005).
8. Ojard, G., Gowayed, Y., Chen, J., Santhosh, U., Ahmad J., Miller, R., and John, R., "Time-Dependent Response of MI SiC/SiC Composites Part 1: Standard Samples", Ceramic Engineering and Science Proceedings. Vol. 28, no. 2, pp. 145-153. 2008
9. Ojard, G., Morscher, G., Gowayed, Y., Santhosh, U., Ahmad J., Miller, R. and John, R., "Thermocouple Interactions During Testing of Melt Infiltrated Ceramic Matrix Composites", Ceramic Engineering and Science Proceedings, pp. 11-20. 2008.
10. Gossner, M., Personal Communication, P&W Fellow, Pratt & Whitney, East Hartford, CT.
11. Du, H., Unpublished Research, United Technologies Research Center, East Hartford, CT.
12. M.R. Jackson and R.L. Mehan, *Ceram. Eng. Sci. Proc.*, vol.2 [7-8], 1981, pp.787-791.

FABRICATION AND PROPERTIES OF HIGH THERMAL CONDUCTIVITY SILICON NITRIDE

You Zhou, Hideki Hyuga, Tatsuki Ohji, and Kiyoshi Hirao
National Institute of Advanced Industrial Science and Technology (AIST)
Nagoya 463-8560, Japan

ABSTRACT

This paper gives an overview on the recent developments of high-thermal-conductivity silicon nitride ceramics. Some major factors lowering thermal conductivity of silicon nitride are clarified and some potential approaches to realize high thermal conductivity are described. Then, the recent achievements on the silicon nitride fabricated by a sintered reaction bonded silicon nitride (SRBSN) process are presented. Due to the reduction of the content of impurity oxygen dissolved in the lattice, the silicon nitride ceramics attained substantially higher thermal conductivities, compared to that of the conventional gas-pressure sintered silicon nitride, while the microstructures and bending strengths are similar to each other between these two types of materials. Moreover, further improvement of the thermal conductivity is possible by increasing the β/α phase ratio of the nitrided compact, and a silicon nitride with a thermal conductivity as high as 177 W/(m·K) and a high fracture toughness of 11.2 MPa·m$^{1/2}$ was developed.

INTRODUCTION

Power electronic devices capable of efficient control and conversion of electric power have been widely used for a variety of applications such as industrial robots, hybrid motor vehicles, and advanced electric trains.[1] In the near future, the replacement of Si semiconductor by the wide-bandgap materials (SiC and GaN) is expected to allow higher voltage, larger current, higher power density, and smaller size of the power devices.[2] Since large electric power is controlled and converted by the power devices, high electric insulation, high heat dissipation, and high heat resistance are required to their circuit substrates. Aluminum nitrides, which possess high thermal conductivity, have been used for circuit substrates of power devices with high power density such as in-vehicle inverters. However, higher power density is strongly demanded nowadays in many applications as above mentioned. In addition, the devices loaded in motor vehicles are subject to large temperature changes. These severe working conditions will lead to generation of cracks in ceramic substrates due to high thermal stresses arising at joints with conductive circuits as shown in Fig. 1.[3] Therefore, both excellent mechanical reliability and high thermal conductivity are required for the substrates used for the high-power devices.

Figure 2 shows relationship between strengths and thermal conductivities of alumina, aluminum nitride, and silicon nitride ceramics commercially available for the substrates currently.[4, 5] While aluminum nitrides have high thermal conductivities, their strengths are low. The thermal conductivities of commercially available silicon nitrides are less than 90 W/(mK), though their strengths are more than twice higher than those of aluminum nitrides. An important issue in silicon nitride is, therefore, to increase thermal conductivity without degradation of mechanical properties. On the other hand, the intrinsic value of the theoretical thermal conductivity of a β-Si$_3$N$_4$ crystal has been estimated to exceed 200 W/(m·K).[6, 7] Thus, it may be possible to prepare silicon nitride ceramics with both good mechanical properties and high thermal conductivities, which would then make them promising substrate materials for the next-generation high-power devices.

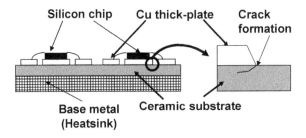

Figure 1. Structure of ceramic substrate for power device.

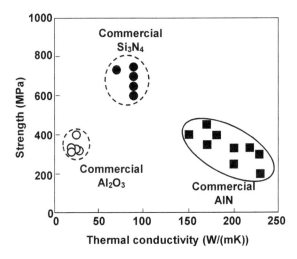

Figure 2. Relationship between strengths and thermal conductivities of commercially available ceramic substrates.[4]

FACTORS LOWERING THERMAL CONDUCTIVITY

Silicon nitride mainly exists in two hexagonal polymorphs, namely α- and β-Si_3N_4, which are generally regarded as low and high temperature crystal forms, respectively. In general, α-Si_3N_4 powders are used as starting raw powder since the α- to β- phase transformation during liquid-phase sintering leads to development of rod-like β-Si_3N_4 grains on account of preferential growth rate in the [001] direction of β-Si_3N_4 crystal. Although the estimated theoretical thermal conductivity of β-Si$_3$N$_4$ crystal is higher than 200 W/(m·K) as above mentioned, the thermal conductivities of commercially available silicon nitride ceramics are much lower for the following reasons.

Due to its strong covalency and low diffusivity, the sinterability of silicon nitride is low. Densification of silicon nitride is usually accomplished by a liquid-sintering mechanism, where some oxide sintering aids are added and they react with silicon nitride as well as the silica phase on the surface of silicon nitride particles to form a liquid phase which promotes densification through rearrangement and solution-reprecipitation mechanisms during heating. After sintering, the liquid phase remains as glassy or partially-crystallized phases in the sintered material. They may exist as isolated secondary phases at the triple point junctions surrounded by three grains, or as continuous thin film (around 1 nm thickness) on the boundaries between two adjacent grains. Because the thermal conductivities of these secondary phases are quite low (1 W/(m·K) or lower), their existence in the microstructure causes reduction of the thermal conductivity of the sintered material. Compared with the triple-junction phases, the detrimental effect of the grain boundary phases is greater due to its continuity. Previous simulation study has revealed that this kind of detrimental effect may be alleviated when the grain sizes of the silicon nitride grains are larger than several micrometers.[8, 9] Thus, promoting grain growth is one of the effective approaches to obtaining a high thermal conductivity.

Besides the secondary and grain boundary phases which reside outside silicon nitride grains, there exist a variety of imperfections called lattice defects (point, line, planer defects) inside the grains. Because heat transfer in silicon nitride occurs by lattice vibration (phonon), lattice defects in the silicon nitride crystals induce phonon scattering and thereby reducing thermal conductivity. The detrimental effect of the lattice defects on the thermal conductivity is greater than those of the above-mentioned secondary and grain boundary phases. It has been reported that solution of oxygen into Si$_3$N$_4$ crystals generates vacancies at the Si sites in the silicon nitride lattice as expressed by the following equation:[10, 11]

$$2SiO_2 \rightarrow 2Si_{Si} + 4O_N + V_{Si} \qquad (1)$$

Where O_N and V_{Si} is a dissolved oxygen atom in a nitrogen site and a vacancy in Si site, respectively. Because the silicon vacancies generated by the dissolution of oxygen in the lattice scatter phonons and reduce thermal conductivity, it is essentially important to lower the lattice oxygen content so as to improve thermal conductivity. The effective approaches to decreasing lattice oxygen content in Si$_3$N$_4$ crystals include using sintering additives with high oxygen affinity (*e.g.* rare earth oxides),[12] and increasing nitrogen/oxygen ratio in the liquid phase.[10] For example, while Yb$_2$O$_3$-MgO sintering additive system resulted in a thermal conductivity of 120 W/(m·K) sintering at 1900 °C for 48 hours, the thermal conductivity increased to 140 W/(m·K) by using MgSiN$_2$ instead of MgO to increase the nitrogen/oxygen ratio.[13] However, the effect is limited if the nitrogen/oxygen ratio is controlled only through selection of sintering additives; the combination of reaction bonding and post sintering that is described in the following section is another effective approach.

HIGH THERMAL CONDUCTIVITY THROUGH REACTION BONDING AND POST SINTERING

The sintered reaction bonded silicon nitride (SRBSN) process is a well-known method of fabricating silicon nitride ceramics from silicon starting powders.[14] In this approach, the whole process from nitridation to post-sintering can be carried out without exposing the compacts to the air, therefore it is favorable for controlling the oxygen content in the obtained silicon nitride sintered body. Furthermore, even a coarse Si powder is used as a starting material, the large Si particles will decompose and transformed to finer Si_3N_4 particles during nitridation, and simultaneously resulting in a nitrided body having a higher relative density (typically ~74%) compared to the Si green body, as shown in Fig. 3. In contrast, when starting from silicon nitride powders, it usually is difficult to attain a green density higher than 60% even if using a CIPing pressure of 400 MPa. The higher density of the nitride body can promote densification and shorten sintering time for the post-sintering process.[15-17]

Figure 3. SEM images of (a) Si raw powder and (b) nitrided compact at 1400°C.[4]

Zhou *et al.* [16] fabricated high-thermal-conductivity silicon nitride ceramics using Si starting powder containing low oxygen content by the SRBSN process. The Si powder had an oxygen content of 0.28 wt%, a total metallic impurity content of <0.01 wt%, and a mean particle size (d_{50}) of 8.5 µm. A combination of 2 mol% Y_2O_3 and 5 mol% MgO were added to the Si powder as sintering additives, and were mixed in methanol using a planetary mill. The collision energy generated by the grinding media induced mechano-chemical reactions, leading to oxidation of the Si particles during the planetary-milling process. Therefore, the oxygen content of the milled Si powders increased to 0.51 wt%, which was still low compared to those of the commercial high-purity Si_3N_4 powders (typically 1 wt% or higher), as shown in Table I.

Figure 4 shows the scanning electron microscopy (SEM) micrographs of the fracture surfaces of the sintered reaction bonded silicon nitride materials which was obtained through reaction-bonding (nitridation) at 1400 °C for 8 hours under 0.1 MPa nitrogen pressure and post-sintering at 1900 °C for various times ranging from 3 to 24 hours under 0.9 MPa nitrogen pressure. They showed some common features: fibrous and faceted grain shapes and a bimodal microstructure where a small fraction of large grains were embedded in a majority of small grains. With increasing sintering time, the microstructures became coarser. For the materials sintered for 3 or 6 h, the microstructures were generally fine, though a few large fibrous grains had lengths of 10 - 20 µm. On the other hand, after being sintered for 12 h, many grains grew to lengths of over 10 µm. For the material sintered for 24 h, a few grains were as long as almost 100 µm.

Table I. Characteristics of the raw Si and Si_3N_4 powders [17]

	Si powder (for SRBSN)	Commercial Si_3N_4 powder (for GPSSN)
Particle size	8.5 μm	0.2 μm
Purity (except oxygen)	> 99.99 %	> 99.9 %
Oxygen content (as received)	0.28 mass%	—
Oxygen content (after milling)	0.51 mass%	—
Oxygen content in Si_3N_4 (after nitridation)*	0.31 mass%	1.2 mass%

*Estimated oxygen content in a fully nitrided material

Figure 4. SEM images of fracture surfaces of SRBSN sintered for various times: (a) 3 h, (b) 6 h, (c) 12 h and (d) 24 h.[16]

Figure 5 shows SEM images of microstructure for SRBSN post-sintered at 1900 °C for 12 hours under 0.9 MPa nitrogen pressure, in comparison with that of gas-pressure sintered silicon nitride (GPSSN) obtained under the same sintering conditions using a commercial high-purity Si_3N_4 powder (mean particle size: 0.2 μm, oxygen content: 1.2 wt%). The four-point bending strengths were 620 MPa and 700 MPa for the SRBSN and GPSSN materials, respectively, indicating no substantial difference between them. However, the thermal conductivity of the SRBSN was 120W/(m·K), which was about 20% higher than that of the GPSSN, 98W/(m·K), due to a smaller amount of impurity oxygen contained in the Si starting powder for the SRBSN material.

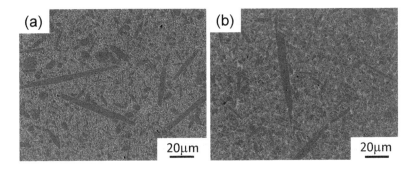

Figure 5. SEM images of microstructures for (a) SRBSN (a) and (b) GPSSN.

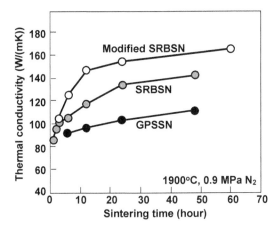

Figure 6. Relationship between strengths and thermal conductivities of GPSSN, SRBSN and modified SRBSN.[17]

Figure 6 shows relationship between sintering time and thermal conductivities of the GPSSN, SRBSN and modified SRBSN materials. With increasing sintering time, thermal conductivities increased for all the samples due to grain coarsening and reducing of lattice oxygen content. It can also be known that the SRBSN materials showed high thermal conductivity compared to the GPSSN materials for all the sintering times. Moreover, Zhou *et al.*[18] investigated the effects of nitridation conditions on the properties of the nitrided and post-sintered samples, and revealed that further improvement of the thermal conductivity was possible by increasing β/α phase ratio of the nitrided sample from conventional 60:40 to 83:17 via controlling the nitrogen atmosphere in the nitridation (the modified SRBSN in Fig. 6). This was thought to be due to further reduced content of impurity oxygen in the post-sintered sample derived from the nitrided sample having high β phase ratio, since it was known that the amount of oxygen which could be solved into β-Si_3N_4 was smaller than that into α-Si_3N_4. The modified SRBSN obtained through sintering at 1900 °C for 60 hours and slow cooling with a rate of 0.2 °C/min demonstrated a further improved thermal conductivity of 177 W/(m·K). This modified SRBSN also showed a very high fracture toughness of 11.2 MPa·m$^{1/2}$ because of the microstructure consisting of large fibrous grains.[18]

Figure 7 shows relationship between thermal conductivities and lattice oxygen contents for the GPSSN, SRBSN and modified SRBSN materials.[17] Despite of the different processing conditions, the data showed a clear correlation that thermal conductivity increased with decreasing lattice oxygen content.

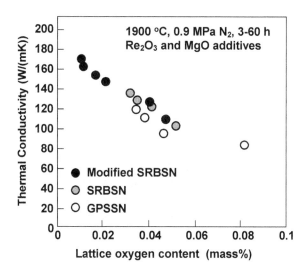

Figure 7. Relationship between thermal conductivities and lattice oxygen contents for GPSSN, SRBSN and modified SRBSN.[17]

SUMMARY

This paper gave an overview on the recent developments of high thermal conductivity silicon nitride ceramics. Some major factors affecting the thermal conductivity of silicon nitride ceramics including secondary phases, grain boundary phases and lattice defects (manly lattice oxygen) were clarified, and some approaches to improving thermal conductivity, such as grain coarsening via long time heat-treatment, using sintering additives with high oxygen affinity, increasing nitrogen/oxygen ratio in liquid phase during sintering were discussed. Following that, our recent studies on preparing high-thermal-conductivity silicon nitride by a sintered reaction bonded silicon nitride (SRBSN) process were reviewed. Because of a smaller amount of impurity oxygen, the SRBSN materials showed higher thermal conductivity compared to the conventional gas-pressure sintered silicon nitride (GPSSN) fabricated from Si_3N_4 starting powder, although the two types of materials had similar microstructures and bending strengths. Moreover, it was found that further improvement of the thermal conductivity was possible by increasing β/α phase ratio of the nitrided sample, resulting in a very high thermal conductivity of 177 W/(m·K) as well as a high fracture toughness of 11.2 MPa·m$^{1/2}$. The concurrent attainment of high thermal conductivity and good mechanical properties of the SRBSN materials will make them promising candidates for the application as insulating substrates for the next-generation high-power devices.

REFERENCES
[1] The Institute of Electrical Engineers of Japan, *Power Semiconductor That Runs the World (in Japanese)*, Ohmsha, Ltd., Tokyo, 2009.
[2] C. R. Eddy Jr. and D. K. Gaskill, "Silicon Carbide as a Platform for Power Electronics," *Science*, **324** [5933] 1398-400 (2009).
[3] M. Yamagiwa, "Packaging Technologies of Power Modules for Hybrid Electric Vehicles and Electric Vehicles (in Japanese)," *Bull. Ceram. Soc. Japan*, 45 [6] 432-37 (2010).
[4] K. Hirao, "Development of Ceramic Substrates with High Thermal Conductivity (in Japanese)," *Bull. Ceram. Soc. Japan*, **45** [6] 444-47 (2010).
[5] K. Hirao, Y. Zhou, H. Hyuga, T. Ohji, and D. Kusano, "High Thermal Conductivity Silicon Nitride Ceramics," *J. Korean Ceram. Soc.*, **49** [4] 380-84 (2012).
[6] J.S. Haggerty and A. Lightfoot, "Opportunities for Enhancing the Thermal Conductivities of SiC and Si_3N_4 Ceramics through Improved Processing," *Ceram. Eng. Sci. Proc.*, **16**, 475-87 (1995).
[7] N. Hirosaki, S. Ogata, C. Kocer, H. Kitagawa, and Y. Nakamura, "Molecular Dynamics Calculation of the Ideal Thermal Conductivity of Single-Crystal α- and β-Si_3N_4," *Phys. Rev. B*, Article No. 134110 (2002).
[8] M. Kitayama, K. Hirao, M. Toriyama, and S. Kanzaki, "Thermal Conductivity of beta-Si_3N_4: I, Effects of Various Microstructural Factors," *J. Am. Ceram. Soc.*, **82** [11] 3105-12 (1999).
[9] A. Okada and K. Hirao, "Conduction Mechanism and Development of High Thermal Conductive Silicon Nitride (in Japanese)," *Bull. Ceram. Soc. Japan*, **39** [3] 172-76 (2004).
[10] M. Kitayama, K. Hirao, A. Tsuge, K. Watari, M. Toriyama, and S. Kanzaki, "Thermal Conductivity of beta-Si_3N_4: II, Effect of Lattice Oxygen," *J. Am. Ceram. Soc.*, **83** [8] 1985-92 (2000).
[11] K. Hirao, K. Watari, H. Hayashi, and M. Kitayama, "High Thermal Conductivity Silicon Nitride Ceramics," *MRS Bull.*, **26** [6] 451-55 (2001).
[12] K. Watari, "High Thermal Conductivity Non-oxide Ceramics," *J. Ceram. Soc. Japan*, **109** [1] S7-S16 (2001).

[13] H. Hayashi, K. Hirao, M. Toriyama, S. Kanzaki, and K. Itatani, ''MgSiN$_2$ Addition as a Means of Increasing the Thermal Conductivity of β Silicon Nitride,'' *J. Am. Ceram. Soc.*, **84** [12] 3060-62 (2001).

[14] A. J. Moulson, ''Reaction-Bonded Silicon Nitride: Its Formation and Properties,'' *J. Mater. Sci.*, **14**, 1017-51 (1979).

[15] X. W. Zhu, Y. Zhou, K. Hirao, and Z. Lences, ''Processing and Thermal Conductivity of Sintered Reaction-Bonded Silicon Nitride. I: Effect of Si Powder Characteristics,'' *J. Am. Ceram. Soc.*, **89** [11] 3331-39 (2006).

[16] Y. Zhou, X. W. Zhu, K. Hirao, and Z. Lences, "Sintered Reaction-Bonded Silicon Nitride with High Thermal Conductivity and High Strength," *Int. J. Appl. Ceram. Technol.*, **5** [2] 119-26 (2008).

[17] Y. Zhou and H. Hyuga, "Development of High Thermal Conductivity Silicon Nitride Ceramics (in Japanese)," *Bull. Ceram. Soc. Japan*, **47** [1] 12-17 (2012).

[18] Y. Zhou, H. Hyuga, D. Kusano, Y. Yoshizawa, and K. Hirao, "A Tough Silicon Nitride Ceramic with High Thermal Conductivity," *Adv. Mater.*, **23** [39] 4563-67 (2011).

POROUS SILICON CARBIDE DERIVED FROM POLYMER BLEND

Ken'ichiro Kita, Naoki Kondo

National Institute of Advanced Industrial Science and Technology (AIST)
2266-98 Shimo-shidami, Moriyama-ku, Nagoya, 463-8560, JAPAN

ABSTRACT

By using polymer blends that contain polycarbosilane and polymethylphenylsiloxane, porous silicon carbide membranes can be obtained by thermal oxidation curing. However, the mechanism of pore construction by thermal oxidation curing can be expected to be different from the mechanism of pore construction by curing by gamma ray oxidation. Modification of curing and pyrolysis methods was found to be effective in the modification of the shape of silicon carbide derived from the polymers.

1. INTRODUCTION

Porous ceramics can be used in various applications such as filtration, heat insulation, absorption, catalysts and catalyst supports. Numerous methods have been reported [1-8] for making porous ceramics. Dacquin et al. [9] prepared porous ceramics by using a mixture containing aluminum isopropoxide, a triblock copolymer, and very small polystyrene beads. By the decomposition of the beads, pores could be generated in the ceramics. In another instance, Fukushima et al. [10] invented the production of ceramics by using a combination of a polymer blend (consisting of two or more kinds of polymers that were well mixed) and the freezing method. Phase separation occurred by freezing and one of the separated phases that could be decomposed easily formed the pores during pyrolysis. As shown in these examples, porous ceramics derived from precursor polymer need a pore generator during pyrolysis.

The investigation on porous silicon carbide fibers with pore diameters of about 20 nm has been previously reported and the fibers were derived from a polymer blend containing polycarbosilane (PCS) and polymethylphenylsiloxane (PMPhS) with curing by gamma ray oxidation [11]. Gamma ray irradiation forced the polymer blend to undergo phase separation and caused strong oxidation; pores were formed by the decomposition and growth of SiC crystals during both the pyrolysis steps [12]. However, the method that uses gamma rays is unfavorable because gamma ray curing is complex to handle during manufacturing of composites owing to the difficulty involved in the regulation of radiation.

To address this challenge, we tried curing with thermal oxidation instead of irradiation to generate pores. Oxidation is causes the formation of pores and sufficient oxidative curing can be expected to substitute for gamma ray curing. However, PMPhS is difficult to be oxidized and it is easy to prevent any other polymer from undergoing oxidation by thermal oxidative curing [13].

In this study, the extension of curing time and the increase of the curing temperature were carried out and the results were compared with that obtained by curing by thermal oxidation of PCS alone. Polysiloxane including phenyl groups begins to decompose at 523 K and we assumed that the decomposition enabled the polymer blend that included the siloxane to be sufficiently oxidized. In addition, we applied this method to porous alumina to produce novel porous ceramics. Obtaining a porous SiC architecture on a large scale is difficult because of the use of polymers, and hence, a ceramic support is necessary. Modification of alumina into silicon carbide has been previously accomplished. Therefore, we expected the production of porous alumina with a porous silicon carbide surface [14].

2. EXPERIMENTAL PROCEDURE

We prepared two kinds of polymer blends containing PCS (NIPUSI-Type A, Nippon Carbon, Japan) and PMPhS (KF-54, Shin-Etsu Chemicals Co. Ltd., Japan). Hereafter, the polymer blends including PCS and PMPhS were referred to as PS polymers. One was the polymer blend that included 15 wt% of PMPhS, which will henceforth be referred to as PS15. The other was the polymer blend including 30 wt% of PMPhS, which will be referred to as PS30. These polymer blends were the same precursor polymers used for the production of porous silicon carbide fiber with a pore diameter of about 20 nm.

The outline of this experiment is shown in Fig. 1. First, a silicon wafer and alumina board with purity of >99.9% were prepared. The wafer was dipped into the PS15 solution and the alumina was dipped into the PS30 solution. Toluene was used as the solvent to dissolve these polymers and the concentration of the polymers was maintained at 0.1 mol/L. After dipping, curing by thermal oxidation was carried out under air flow. During curing, the heating rate was fixed at 4 K/h up to 523 K and the temperature was maintained at 523 K for 1 h. Then, the sample was pyrolyzed at 1273 K for 1 h under the flow of Ar and then re-pyrolyzed at 1623 K for 1 h under the flow of Ar. To avoid melting of the silicon wafer, the temperature of re-pyrolysis was reduced when compared to the temperature used in a previously reported experiment [15]. The mass residues after curing, after pyrolysis at 1273 K, and re-pyrolysis at 1623 K were measured. In the case of PS30 on alumina, the sample re-pyrolyzed at 1673 K instead of 1623 K was also prepared, and the effect of Ar flow during thermal decomposition was investigated. After the re-pyrolysis, the samples were observed by scanning electronic microscopy (SEM; JEM-5600, JEOL, Japan) and X-ray diffraction (XRD; RINT2500, Rigaku Corporation, Japan)

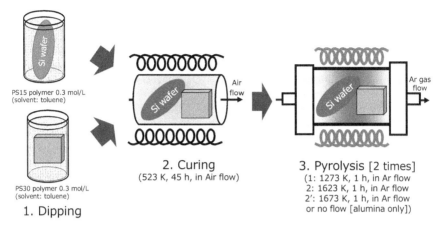

Figure 1. Schematic describing the outline of the experiment carried out in this study.

3. RESULTS AND DISCUSSION

A previous study had concluded that the increase in mass residue during thermal oxidation was connected with the formation of tiny pores [11]. Therefore, analyzing the mass residues during the presently carried out experiments was necessary to investigate the possibility of pore formation. Fig. 2 shows the mass residues obtained during the experiments carried out and the results of curing by gamma ray oxidation derived from a former investigation for comparison. It is noteworthy that while the mass gain of PS15 and PS30 after curing by gamma ray oxidation was approximately 10 wt%, the mass gain of PS15 and PS30 after curing by thermal oxidation was approximately 1 wt%. Curing by thermal oxidation at 523 K seemed to fail to induce oxidation of PS15 and PS30 because of the strong resistance to oxidation derived from the presence of PMPhS. However, the mass residue of PS15 and PS30 after re-pyrolysis at 1623 K was approximately 60 wt% and this value was almost identical to that of PS30 samples obtained after curing by gamma ray oxidation and re-pyrolysis at 1623 K. The mass decrease can be ascribed to the thermal decomposition of the samples, revealing that thermal decomposition occurred despite poor curing by thermal oxidation.

Fig. 3 shows the XRD patterns of the various samples and the pattern acquired from the porous SiC fiber obtained by curing by gamma ray oxidation and subsequent pyrolysis at 1673 K is shown for comparison. After pyrolysis at 1673 K, these polymers transformed into amorphous SiC and continuous peaks appeared at around 18°, 36.5°, and 60°. The polymers on silicon wafer or alumina board used in this study showed these peaks at around 18° and 35°, confirming the presence of the   polymers on silicon wafer or alumina board. However, these peaks were very weak, which could be ascribed to the excessively high angle of the incoming beam. The detailed investigation of the XRD patterns showed the presence of amorphous SiC. In addition, the pattern acquired from the PS30 sample pyrolyzed at 1673 K in the absence of Ar flow showed the peaks of aluminum silicate and silica.

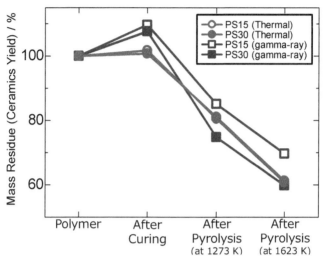

Figure 2.   Mass residues of PS15 and PS30 polymers after curing and pyrolysis.

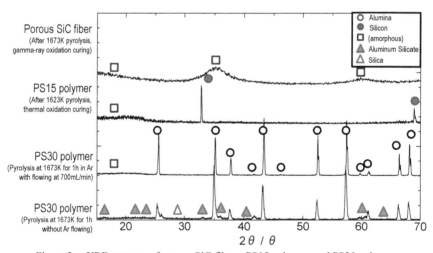

Figure 3.   XRD patterns of porous SiC fibers, PS15 polymer, and PS30 polymers.

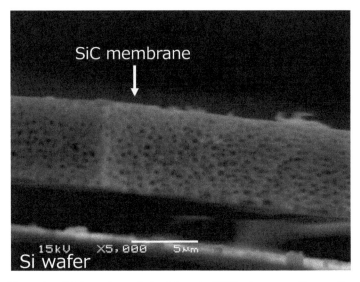

Figure 4.   SiOC membrane derived from the PS15 polymer on silicon wafer.

Figure 5.   SiOC membrane derived from the PS30 polymer on alumina board
after pyrolysis at 1673 K with Ar flow.

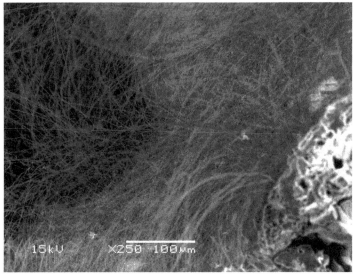

Figure 6.   SiOC membrane derived from the PS30 polymer on alumina board after pyrolysis at
1673 K without Ar flow.

Fig. 4 shows the silicon oxycarbide (SiOC) membrane derived from the PS15 polymer on silicon wafer. This membrane was transformed into porous SiOC membrane and the pore diameter was about 600 nm. It is to be noted that although the uniformity of the pore shape was identical to that of the porous SiC derived from PS15 subjected to curing with gamma rays, the pore diameter was excessively large when compared to that of the porous SiC obtained by gamma ray curing (the pore sizes were about 20 nm). Fig. 5 shows the SiOC membrane derived from the PS30 polymer on alumina board. The membrane was also transformed into a porous SiOC membrane. However, the pore diameter was more than 1 μm i.e., larger than that obtained in the case of the PS15 polymer on silicon wafer. Besides, this membrane seemed to melt during pyrolysis, which can be attributed to the lack of curing and hence, the curing temperature and time were insufficient for the PS30 polymer. Fig. 6 shows the SiOC membrane derived from the PS30 polymer on alumina board in the absence of Ar flow. It can be seen that the membrane was entirely transformed into nanowires. Considering the XRD pattern acquired and those obtained in a former study, the nanowires were found to be made of silica and were generated by the retention of gases including silane, carbon mono-oxide etc. obtained from the decomposition of the polymer [16] Aluminum silicate might have been derived from the boundary between the polymer and alumina because aluminum silicate has always been generated during SiC modification of alumina surface [17].

4. CONCLUSION

In this study, we noticed some differences between PS polymers oxidized by gamma rays and the polymers by thermal curing. The mass increase by thermal curing was less than that caused by gamma ray curing. The mass residue obtained after pyrolysis at 1673 K was almost identical to that obtained with PS30 subjected to curing by gamma ray oxidation. The pore diameter of the SiC membrane obtained in this study was larger than that obtained with irradiation. However, the structure of the pores observed in this study did not seem to be connected to the thermal decomposition of the uniform Si-O-C layer.

REFERENCES

[1] P. Colombo, and M. Modesti, "Silicon Oxycarbide Ceramic Foams from a Preceramic Polymer", *J. Am. Ceram. Soc.*, **82** (1999), 573-578.

[2] Y. –W. Kim, K. –H. Lee, S. –H. Lee, and C. B. Park, "Fabrication of Porpus Dilicon Oxycarbide Ceramics by Foaming Polymer Liquid and Compression Molding", *J. Ceram. Soc. Jpn.*, **111** (2003), 863-864.

[3] J. Zeschky, T. Höfner, C. Arnold, R. Weißmann, D. Bahloul-Hourlier, M. Scheffler, P. Greil, "Polysilsesquioxane derived ceramic foams with gradient porosity", *Acta Mater.*, **53** (2005), 927-937.

[4] E. Bernardo, "Micro- and macro-cellular sintered glass-ceramics from wastes" *J. Euro. Ceram. Soc.*, **27** (2007), 2415-2422.

[5] Q. Wang., G. –Q. Jin, D. –H. Wang, and X. –Y. Guo, "Biomorphic porous silicon carbide prepared from carbonized millet", *Mater. Sci. Eng. A*, **459** (2007), 1-6.

[6] S. Shiraishi, A. Kikuchi, M. Sugimoto, and M. Yoshikawa, "Preparation of Silicon Carbide-based Nanoporous Materials by Replica Technique", *Chem. Lett.*, **37** (2008), 574-575.

[7] B. V. M. Kumar, W. Zhai, J. –H. Eom, Y. –W. Kim, and C. B. Park, "Processing highly porous SiC ceramics using poly(ether-co-octene) and hollow microsphere templates", *J. Mater. Sci.*, **46** (2011), 3664-3667.

[8] T. Ohji and M. Fukushima, "Macro-porous ceramics: processing andproperties," *Inter. Mater. Rev.*, **57** (2012), 115-131

[9] J. -P. Dacquin, J. Dhainaut, D. Duprez, S. Royer, A. F. Lee, and K. Wilson "An Efficient Route to Highly Organized, Tunable Macroporous-Mesoporous Alumina", *J. Am. Chem. Soc.*, **131** (2009), 12896-12897.

[10] M. Fukushima, M. Nakata, Y. Zhou, T. Ohji, and Y. Yoshizawa, "Fabrication and properties

of ultra highly porous silicon carbide by the gelation-freezing method", *J. Euro. Ceram. Soc.*, **30** (2010), 2889-2896.

[11] K. Kita, M. Narisawa, A. Nakahira, H. Mabuchi, M. Sugimoto, and M. Yoshikawa, "Synthesis and properties of ceramic fibers from polycarbosilane/polymethylphenylsiloxane polymer blends", *J. Mater. Sci.*, **45** (2010), 3397-3404.

[12] K. Kita, M. Narisawa, H. Mabuchi, M. Itoh, M. Sugimoto, M. Yoshikawa, "Synthesis of SiC Based Fibers with Continuous Pore Structure by Melt-Spinning and Controlled Curing Method", *Adv. Mater. Sci.*, **66** (2009), 5-8.

[13] W. Noll, *"Chemistry and Technology of Silicones"*, (Waltham, MA: Elsevier, 1968), 409-426.

[14] K. Kita, N. Kondo, Y. Izutsu, H. Kita, "Investigation of the properties of SiC membrane on alumina by using polycarbosilane", *Mater. Lett.*, **75** (2012), 134-136.

[15] K. Kita, M. Narisawa, A. Nakahira, H. Mabuchi, M. Itoh, M. Sugimoto, M. Yoshikawa, "High-temperature pyrolysis of ceramic fibers derived from polycarbosilane–polymethylhydrosiloxane polymer blends with porous structures", *J. Mater. Sci.*, **45** (2010), 139-145.

[16] Y. Hasegawa, and K. Okamura, "Synthesis of continuous silicon carbide fibre Part 3 Pyrolysis process of polycarbosilane and structure of the products", *J. Mater. Sci.*, **18** (1983), 3633-3648.

[17] K. Kita, N. Kondo, H. Hyuga, Y. Izutsu, H. Kita, "Study of modification on alumina surface by using of organosilicon polymer", *J. Ceram. Soc. Jpn.*, **119** (2011), 378-381.

PROCESSING AND PROPERTIES OF ZIRCONIA TOUGHENED WC-BASED CERMETS

I. Hussainova[1], N. Voltsihhin[1], M. E. Cura[2], S-P. Hannula[2]

[1]Department of Materials Engineering, Tallinn University of Technology,
Ehitajate 5, 19086 Tallinn, Estonia;
[2]Department of Materials Science and Engineering, Aalto University, P.O.Box 16200,
FI-00076 AALTO Finland

ABSTRACT

The outstanding toughening capability exhibited by zirconia-based ceramics has raised considerable attention in materials doped by tetragonal zirconia. To obtain materials of desired microstructure and provide properties required for the different applications manufacturing process should be considered in details. The main aim of this work was to prototype the composite characterized by the uniformly dispersed zirconia grains of the sub-critical size throughout the matrix of WC-Ni and to retain tetragonal polymorphs of ZrO_2 during sintering. In the present study, cermets were consolidated with the help of spark plasma sintering technique and their microstructure and mechanical properties were evaluated.

INTRODUCTION

Among many hard alloys available on the market, cemented tungsten carbides find a wide range of industrial applications due to their unique combination of mechanical, physical and chemical properties. Most of the WC-Co applications, starting from cutting tools through dies to teeth on gravel extractors, require high wear resistance and tolerance to impact loading. The need for hardmetals with improved properties, particularly fracture toughness and corrosion resistance, has attracted research interests on development of fine-grained tough composites [1-5].

To increase a corrosion and oxidation resistance of usually used cobalt binder, the focus is made on nickel binder in WC-Ni hardmetals.[6,7]Attempts to incorporate zirconium dioxide into the WC matrix to substitute binder metal has recently meet with the success in fabrication of WC-ZrO_2 composites of superior properties and excellent chemical stability.[1-5] Along with high flexural strength of about 1 GPa and fracture toughness of about 10 MPa m$^{1/2}$, zirconia also possesses a relatively high hardness, wear and thermal shock resistance [8, 9]. Alloying with suitable aliovalent cations stabilizes tetragonal zirconia at room temperature and improves its functional properties. Moreover, utilization of zirconia as dopant particles is of special interest because of chemical stability at high temperatures and possible stress-induced martensitic phase transformation from tetragonal to monoclinic modification (t-m) during crack propagation [1, 5, 8].

To enhance materials performance under mechanical and/or tribological loading, fine –grained structures are required. To inhibit grain growth during sintering, two main approaches are widely used: (i) exploitation of rapid sintering methods such as, for example, spark or electric field assisted sintering (SPS); and (ii) incorporation of grain growth inhibitors to mitigate grain growth during sintering [4, 7]. From a phenomenological point of view SPS is similar to a hot pressing due to consolidation by the simultaneous application of mechanical pressure and heating. However, application of a high electric current for SPS allows consolidating powders to full density much faster and at lower temperatures as compared to conventional methods of powder metallurgy. In this work the SPS route and yttria stabilized zirconia (YSZ) as a possible grain growth inhibitor were utilized for WC-YSZ-Ni fabrication.

From a practical point of view, the essential requirement to the processing route is obtaining the uniform distribution of phases throughout a composite and a dense material with low residual flaw population to ensure contribution of zirconia transformation into toughening process. Therefore, the main aim of the present study was to prototype a reliable hardmetal exploiting a synergistic effect of grain refinement; transformation toughening and crack bridging by ductile constituent in order to obtain structurally efficient composite.

EXPERIMENTAL
WC powder with average particles size of 0.9 μm and 99% of purity was provided by Wolfram GmbH, Austria. The commercially available nickel powder of 99.7% purity and average grain size of 20 μm was used as a binder metal. 6 wt% of a commercial 3 mol% yttria stabilized tetragonal zirconia powder (< 100 nm, 99.9 % purity, Alfa Aesar, Germany) was mixed with the precursor powders to produce the WC − YSZ − Ni hardmetal. To eliminate unfavorable η-phase formation during sintering 0.2 wt.% of free carbon was added to the mixture. Therefore, the final composition of the blend is 85.8wt.%WC − 6wt.%YSZ − 8wt.%Ni − 0.2wt.%C or, considering tetragonal zirconia density as 6.05 g cm^{-3}, 73.22vol.% WC − 12.12vol.% YSZ − 13.37vol.%Ni − 1.28vol.%C. The optimal amount of carbon to be added has been estimated by comparing the final microstructures obtained from the mixtures with different amount of C as outlined in the authors' work [10].

The powders were mixed in ethanol with 3g of PEG (Polyethylene glycol) in a rotary ball mill for 72 hours using 6 mm diameter WC-Co milling balls. WC powder was subjected to a high energy milling to reduce the particle size to sub-micron. After milling the component powders were mixed with the help of a low energy mixing mode.

Milled and dried powder blends were subjected to a cold pressing at 8.5 MPa to obtain green compacts with a green density of about 55% of theoretical density. Green bodies have been held at 500 ∘C in hydrogen for 30 min to burn plasticizers off.

X-ray diffraction (XRD; Siemens, Bruker D5005) analysis was conducted to identify phases in the initial powder and crystalline phases present after sintering. Specimens were irradiated with Cu K_α - radiation at 40 kV in a scanning range 2θ from 20° to 80° with a step size of 0.04°. Microstructural imaging was performed using scanning electron microscopy (SEM; Zeiss EVO MA15 supplied with energy dispersive microscope EDS - INCA analyzer).

During the consolidation of the powders by SPS method a FCT HP D 25-2 (FCT, Germany) unit was used. The powders were compacted in cylindrical graphite moulds with an inner diameter of 20.8 mm while graphite foils with a thickness of 0.4 mm were placed in between the powder and the graphite surfaces to increase the contact area between them. Graphite felt was used around the mould for insulation and avoiding the thermal gradient. Retention of the metastable tetragonal polymorph of zirconia is critically dependent on the sintering conditions and affected by the sintering temperature to a great extent. During sintering the processing temperature was measured by a pyrometer through a hole in the upper punch from the thin graphite wall surface at 5 mm distance from the powder. The current was applied directly to the material from the upper electrode through upper punch in a pulse pause ratio of 10ms:5ms. The sintering experiments were carried out under a vacuum of 50–60 mTorr at temperatures of 1200 °C under a pressure of 50 MPa. The maximum pressure was applied at the room temperature and kept on the material until the start of cooling and removed gradually as the temperature reached 25 °C.

Prior to the investigations, the specimens (25 mm × 15 mm × 5 mm in size) were grounded and smoothly polished with diamond paste to obtain optically reflecting surfaces.

The microstructure and grain size of the constituent phases were examined with scanning electron microscopy (SEM, Leo Supra-35) equipped with energy dispersive spectroscopy (EDS − TM-1000) analyzer. Grain size was described by the spherical equivalent diameter. X-ray diffraction (Siemens Bruker D5005X-ray analyzer with Cu Kα-radiation and scanning range 2θ from 20^0 to 80^0 with a step of 0.04^0) was used to perform a quantitative analysis of the phases on polished as well as fractured surfaces. The integrated intensities of the (101) tetragonal peak and ($\bar{1}$11) and (111) monoclinic peaks were measured and volume fraction of the monoclinic polymorph was calculated using the Porter and Heuer method corrected for the monoclinic – tetragonal system[11].

The sintered density was measured by the Archimedes technique using water as the immersion medium. Hardness and modulus of elasticity measurements were conducted according to EN ISO 14577 on a ZWICK tester (ZWICK, Ulm, Germany) applying an indentation load of 10 kg. The Vickers hardness value for each composition was taken as the average of 10 indents. The indentation toughness approach was chosen as the easiest and effective way to estimate fracture toughness of the composite. The indentation fracture toughness (K_{IC}) was calculated from the length of radial cracks emanating from the corners of the Vickers indentation imprints using the well-known Palmqvist approach. Load of 50 kg on the indenter resulted in quite well developed cracks that would mitigate any surface effects of the indentation. The reported values are the mean and standard deviation of at least six indentations.

RESULTS AND DISCUSSION

Microstructure
The XRD pattern as well as a SEM image of the mixture is shown in Figure 1. Images in Figure 2 present the microstructure of a specimen produced.

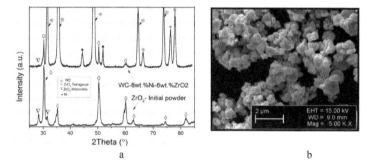

a   b

Figure 1. (a) XRD patterns of the initial zirconia powder and precursor powder blend; and (b) SEM micrograph of the precursor WC powder.

The shape of carbide grains in conventional WC–Co hardmetals is usually a truncated triangular prism and influenced by the binder metals and/or additives. In the composites developed during this study, three types of the WC grains have been detected: small amount of standard triangular shaped grains; some amount of rectangular shaped grains; while most of the WC grains were relatively equiaxial. The microstructure can be considered as an ultra-fine grained one with the WC grain size in the sub-micron range of 0.5 µm, which is about the particle size of the precursor WC powder after milling/mixing process (0.1 – 0.6 µm). The boundaries of the WC grains cannot be clearly determined from the SEM images, but well developed structure is indisputable. This result implies minimal grain growth during processing. Therefore, it may be concluded that rapid sintering process and possibly dispersion of ZrO_2 particles are useful in suppressing the WC grain growth. The effect of oxides on grain growth is still quite questionable and requires further evaluation.

Zirconia grains can also be determined as ultra-fine sized although zirconia particles are highly agglomerated and present clusters of quite uniform size distribution ranging between 0.05 and 0.5 μm and most of them are located between fine carbide grains surrounded by binder metal.

XRD patterns recorded from the polished surfaces of the processed sample are presented in Figure 3. Within the limits of XRD detection, no phases other than WC, tetragonal zirconia, and Ni were recorded from the corresponding patterns. This suggests that no sintering reaction occurred to any noticeable extent during the processing. Another important observation is the presence of tetragonal zirconia. This suggests that yttria stabilized ZrO_2 is retained in its metastable tetragonal polymorph in the present processing scheme and, therefore, can contribute into increase in fracture toughness of the composite.

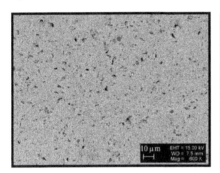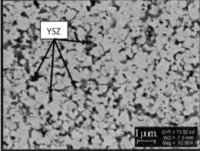

Figure 2. SEM images of the WC – YSZ – Ni hardmetal at different magnification.

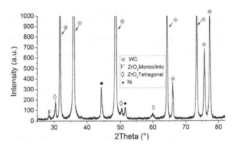

Figure 3. XRD patterns for hardmetals produced SPS route.

The volume change accompanying transformation creates a compressive strain field around a crack tip to oppose crack propagation, while the strain energy associated with any net shear component of the transformation strain in the transformation zone contributes an effective increase in the energy of fracture[8]. Additional contributions to toughness may result from microcracking associated with accommodation of the transformation shape strain and from crack deflection within the transformation zone ahead of the crack.

The martensitic transformation from the tetragonal to monoclinic structure is strongly influenced by zirconia grain size and size distribution throughout a composite [8, 12, 13].

A thermodynamic formulation for the critical size of inclusions to be transformed in a constrained matrix d_c can be expressed through the process of "competition" in a change of a surface free energy and a chemical free energy during martensitic t $\rightarrow$ m transformation of the zirconia particle. The thorough theoretical consideration of the problem is out of scope of the present paper, however, rough analysis performed by the authors in [12] by comparison of Gibbs energies of the composite materials with inclusions in different phase states gives an estimation of the critical radius of the 3 mol% yttria stabilized zirconia particles to be around 1 μm. The predominant role of the chemical energy changes, as compared to a potential deformation energy density caused by thermo-residual stresses and eigenstrain of phase transformation at room temperature, leads to the fact that the elastic interaction of the inclusions insignificantly influences the critical radius. When the concentration of inclusions increases, the critical radius of particles embedded into WC matrix only slightly increases. In the composites produced, the average zirconia grain size was about 0.3 μm. The narrow size distribution around the critical grain size leads to more effective realization of the mechanism of phase transformation. However, the yttria stabilized zirconia grains are considered to be quite small for effective contribution of the transformation toughening into overall toughness of the material.

Mechanical properties

Hardness and fracture toughness are the two of the most important mechanical characteristics of tribo-materials. Other characteristics such as wear resistance are more or less affected by these two parameters. The hardness (HV_{10}) and indentation fracture toughness of the composite tested are 1705 ± 36 and 11.35 ± 3.3 MPa m$^{1/2}$, respectively. As it was expected, quite high Vickers hardness was achieved through the grain size refinement. Conventional hardmetals of 8 – 12 wt.% of binder metal (Co or Ni) possess hardness in a very wide range starting from as low as about HV_{10} = 1000 and achieving as high value as 1600 depending on many factors including carbide grain size, porosity, and so on [14]. It is well known that the hardness of the ultra-fine grained materials is significantly higher than of coarse grained ones and increases with decrease in WC grains size.

Indentation fracture toughness of the composites developed is somewhat higher as compared to published data for conventional hardmetals with the same amount of metal binder. It has been shown that the crack does not advance exactly along the interfaces in conventional WC-Co/Ni hardmetals but proceeds in the binder, forming closely spaced shallow dimples in material. There is an appreciable amount of plastic deformation of the binder involved into the process of fracture. The increase in the fractions of crack path through binder – ceramic grain interfaces in ultra-fine hardmetals will contribute a significant amount of fracture energy to the process, which would in turn enhance the overall toughness of the material. For conventional bulk WC–Co/Ni, the fracture toughness is a function of mean free path λ between the WC grains. It is dependent on the plastic deformation and tearing of metal binder. A finer grain size usually results in a smaller λ at a given constant volume fraction of binder and a smaller plastic zone, and therefore, lower fracture toughness. For ultra-fine grained hardmetals the effect of the plastic mechanisms of binder metal will no doubt be reduced [15]. The mean free path is estimated to be about 120 nm in the composite tested because of quite high continuity between both WC grains and WC/YSZ grains, which was estimated of about 0.68 for the material. The degree of contiguity greatly influences the properties of composites, particularly if the properties of the constituent phases differ significantly. Generally, as the contiguity increases, hardness increases and fracture toughness decreases. Therefore, some increase in fracture toughness can be expected from synergetic effect of the crack bridging/plasticity of the metal binder as well as the stress activated phase transformation.

However, it should be noticed that because the fracture toughness value was measured indirectly using the Palmqvist method and converted into the K_{IC} values, it is not evident that these values can be reproduced using the standard fracture toughness testing methods such as the short-rod method (ASTM-B771) or the SENB method (ASTM-E399).

Figure 4 shows the SEM micrographs of the indentation cracks propagating in the composite. In all probability the advanced crack transforms the zirconia particles occurring in the path of crack leaving behind wake regions of permanently transformed material. The crack propagates predominantly along grain boundaries indicating an intergranular character of cracking; zirconia grains also tend to deflect the crack. Further, when the crack propagates along the tetragonal grains, some crack closure can be observed pointing to the possible grain transformation due to the crack stress field. Therefore, the multiplicative interactions such as crack deflection and stopping combined with binder plasticity are evident and can be considered as operating toughening mechanisms in the composites.

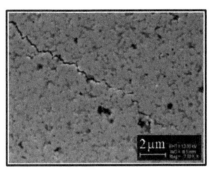

Figure 4. SEM micrograph of the indentation crack path.

CONCLUSION

Ultra-fine grained tungsten carbide based and yttria stabilized zirconia doped hardmetals of high hardness and relatively high fracture toughness can be successfully produced by spark plasma sintering routine.

ACKNOWLEDGEMENTS

The authors would like to acknowledge MS R. Traksmaa and PhD M. Viljus from Materials Research Centre, TUT, for the help with XRD and SEM analysis. Estonian Science Foundation under grant No. 8211 and Estonian Ministry of Education and Research under target financing (project SF 0140062s08) as well as Science Academy of Finland (grant N. 259596) are deeply appreciated for the financial support of this study.

REFERENCES

[1] O. Malek, B. Lauwers, Y. Perez, P. De Baets, J. Vleugels, Processing of ultrafine ZrO2 toughened WC composites. *J. European Ceramic Society*, **29**, 3371–78 (2009)

[2] I. Hussainova, M. Antonov, A. Zikin, Erosive wear of advanced composites based on WC. *Trib. Int.*, **46**, 254 – 60 (2012)

[3] Jiang, D., der Biest, O., Vleugels, J. ZrO_2-WC nanocomposites with superior properties. *J. European Cer. Soc.*, **27**, 1247-51 (2007)

[4] K. Biswas, A. Mukhopadhyay, B. Basu, K. Chattopadhyay, Densification and microstructure development in spark plasma sintered WC–6 wt% ZrO2 nanocomposites, *J. Mater. Res.* **22** 1491–1501 (2007)

[5] Hussainova, I.; Antonov, M.; Voltsihhin, N. Assessment of zirconia doped hardmetals as tribomaterials. *Wear*, **271**, 1909 –15 (2011)

[6] H-C. Kim, I-J. Shon, J-K. Yoon, et al. Rapid sintering of ultrafine WC-Ni cermets, *Int. J. Refractory Met & Hard Mat*, **24**, 427-31 (2006)

[7] H. Rong, Z. Peng, X. Ren, Y. Peng, et al. Ultrafine WC-Ni cemented carbides fabricated by spark plasma sintering, Mat. Sci. Eng. A, 532, 543-7 (2012)

[8] R.H.J. Hannink, P.M. Kelly, B. C. Muddle, Transformation Toughening in Zirconia-Containing Ceramics. *J. Am. Ceram. Soc.,* **83** [3], 461–87 (2000)

[9] Basu, B., Venkateswaran, T., Sarkar, D. Pressureless sintering and tribological properties of WC–ZrO2 composites. *Journal of the European Ceramic Society,* **25(9)**, 1603-1610 (2005).

[10] Voltsihhin, N.; Hussainova, I.; Cura, E.; Hannula S.-P.; Traksmaa, R. Densification and microstructure development in zirconia toughened hardmetals. Key Engineering Materials, **527**, 50-55 (2012)

[11] H. Toraya, M. Yoshimura, S. Somiya. Calibration curve for quantitative analysis of the monoclinic – tetragonal ZrO2 system by X – ray diffraction. *J. Am. Ceram. Soc.*, **67**, 6, C-119 (1984)

[12] A. Freidin, R. Filippov, I. Hussainova, E. Vilchevskaja, Critical radius in the effect of transformation toughening of zirconia doped ceramics and cermets. Key Engineering Materials, **527**, 68 -73 (2012)

[13] A. Suresh, M.J. Mayo, Crystalline and grain size-dependent phase transformations in yttria-doped zirconia. *J. Am. Ceram. Soc.*, **86**, 2, 360 – 2 (2003)

[14] Hussainova, I Microstructure and erosive wear in ceramic-based composites. *Wear*, **258**, 357 – 65 (2005)

[15] Z. Zak Fang, X. Wang, T. Ryu, K.S. Hwang, H.Y. Sohn. Synthesis, sintering and mechanical properties of nanocrystalline cemented tungsten carbide. *Int. J. Refract. Met.& Hard Mat.* **27**, 288 – 99 (2009)

MECHANISM OF THE CARBOTHERMAL SYNTHESIS OF $MgAl_2O_4$-SiC REFRACTORY COMPOSITE POWDERS BY FORSTERITE, ALUMINA AND CARBON BLACK

Hongxi Zhu, Hongjuan Duan, Wenjie Yuan, Chengji Deng
The State Key Laboratory Breeding Base of Refractories and Ceramics, Wuhan University of Science and Technology
Wuhan, Hubei, P.R. China

ABSTRACT

$MgAl_2O_4$ and SiC play an important role in the carbon-containing refractory. In this paper, $MgAl_2O_4$-SiC refractory composite powders were prepared by using forsterite, alumina and carbon black by carbothermal reaction. The effects of the calcined temperature and the proportion of carbon black on phase composition and microstructure of $MgAl_2O_4$-SiC powders were investigated. The mechanism of synthesis processing was discussed. The result shows that $MgAl_2O_4$ crystallite formed by solid phase reaction, most like the reaction of the forsterite and alumina, and its growth occurred by solid-liquid phase reaction. The silicon carbide was synthetized using the vapour-liquid-solid mechanism in a FeSi droplet.

INTRODUCTION

Carbon-containing refractory which is widely used in converter, electric furnace, ladle, and continuous casting so far, has excellent slag penetration and corrosion resistance.[1] These properties would be enhanced, if some good oxidant (such as $MgAl_2O_4$) was added to its.[2,3]

$MgAl_2O_4$ has been recognized as an excellent material for its attractive properties, such as high melting point (2135°C), good mechanical strength, low thermal expansion coefficient, high thermal shock resistance, high chemical attack resistance, excellent optical and low dielectric constant.[4-6]

In order to reduce the oxidation of carbon, antioxidant for metal, alloy and carbide were added to carbon-containing refractory.[7,8] Silicon carbide was the effective antioxidant.[9]

The traditional carbon-containing refractory prepared by powder composed of spinel and silicon carbide after mechanical mixing. $MgAl_2O_4$-SiC powders obtained by carbothermic reaction of forsterite have been studied.[10-13]

In this study, the mechanism of the synthesis of $MgAl_2O_4$-SiC refractory composite powders by carbothermal reduction process using forsterite, industrial alumina and carbon black as starting materials under argon atmosphere were studied.

EXPERIMENTAL PROCEDURE

Natural forsterite (325 mesh, Yichang, China), industrial alumina (200 mesh, purity 94.12 wt.%) and carbon black (800 mesh, purity 96.62 wt.%) were used as raw materials. The chemical composition of forsterite was shown in Table 1, mineral composition of which was mainly forsterite, quartz and hematite.

Table 1 Chemical composition of forsterite (wt.%)

SiO$_2$	Al$_2$O$_3$	Fe$_2$O$_3$	CaO	MgO	K$_2$O	Na$_2$O	TiO$_2$	IL
41.03	0.71	8.67	0.41	46.01	0.022	0.28	0.019	1.37

The carbothermal reduction reactions in forsterite, alumina and carbon black were considered to proceed as follows.

$$Mg_2SiO_{4(s)} +3C_{(s)} +Al_2O_{3(S)} \rightarrow 2MgAl_2O_{4(s)} +SiC(s) +2CO_{(g)} \quad (1)$$

The change in the Gibbs free energy (ΔG) of Eq. 1 is -77.0 kJ/mol at 1923K, -12.617 kJ/mol at 1673 K and 84.32kJ/mol at 1473 K, respectively, which shows that a higher calcination temperature promoted the process of reaction. The thermodynamic calculation value of Gibbs free energy of the reaction was negative when the temperature is above 1708K.The compositions of the different mixes are demonstrated in Table 2, marked as MA1 and MA2.

Table 2 Ratio of forsterite, carbon black and alumina in different samples

sample	Mg$_2$SiO$_4$/C/ Al$_2$O$_3$						forsterite /C/Al$_2$O$_3$		
	molar			mass			mass		
MA1	1	0	1	35	9	0	44	9	0
MA2	1	3	1	35	9	51	44	9	51

All raw materials were mixed at room temperature. The mixture was pressed into specimens with the size of Φ20×20 mm. Then specimens were calcined at 1200 °C, 1400 °C, 1650 °C for 3 hrs in the air or at argon atmosphere at the gas flow velocity 100 ml/min. The phase composition and microstructure of the specimens were analyzed and observed by X-ray powder diffraction (XRD, Philips, X'Pert Pro MPD), scanning electron microscopy (SEM, FEI, Nova 400 Nano) and energy dispersive X-ray spectroscopy (EDS, Oxford Penta FET×3).

RESULTS AND DISCUSSION

Fig. 1 shows the XRD patterns of sample MA1 heated in the air at 1200 °C, 1400 °C, and 1650 °C for 3hrs. When the temperature was 1200 °C, forsterite was reacted with alumina and a small amount of spinel was generated. The phases of forsterite and alumina still existed. At 1400 °C, forsterite was almost completely reacted with alumina, which transformed into spinel and cordierite. When the temperature was 1650 °C, only spinel left. The presence of amorphous phases of the Si-Al-O-Mg system does not show, the reason was amorphous phases are not clearly by XRD. The microstructure of sample MA1 were demonstrated in Fig. 2. The bulk density of samples increased in heating temperature (Fig. 2a at 1200 °C, Fig.2b at 1400 °C, Fig.2c at 1650 °C). The density were improved by the formation of spinel and cordierite phase at 1400 °C. At 1650 °C, higher than the melting point of cordierite (1450 °C), the amorphous phases formed liquid and filled in spaces among the spinal leading to dense the matrix. In Fig. 3, SEM micrograph of sample MA1 in the air at 1400 °C and 1650 °C for 3hrs in the air showed the growth of MgAl$_2$O$_4$ (Fig.3a at 1400 °C, Fig.3b at 1650 °C). The increasing of heating temperature was accompanied by a substantial increase in average MgAl$_2$O$_4$ crystallite size. It was considered that the liquid phase, which produced by the low melting of amorphous phases, benefited for the growth of MgAl$_2$O$_4$. It denoted that MgAl$_2$O$_4$ crystallite formed by solid phase reaction, most like the reaction of the forsterite and alumina, its growth occurred by solid-liquid phase reaction.

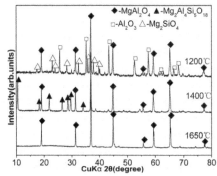

Fig. 1 XRD patterns of MA1 heated at different temperature for 3hrs

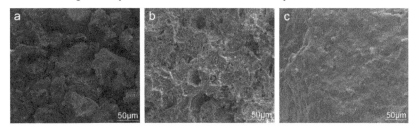

Fig. 2 SEM micrographs of MA1 sample at different temperature

(a) 1200 °C, (b) 1400 °C, (c) 1650 °C.

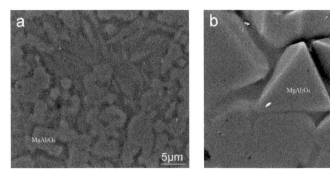

Fig. 3 SEM micrograph of sample MA1 in the air
(a) 1400 °C, (b) 1650 °C

XRD patterns of sample MA2 heated under argon atmosphere at 1200 °C, 1400 °C, and 1650 °C for 3hrs is shown in Fig. 4. The XRD results revealed that no reaction occurred between forsterite and alumina at 1200 °C. Significantly, Fe_2O_3 in the forsterite was reduced, resulting in the presence of silicon iron. The formation of spinel, cordierite and silicon carbide phase were evident at 1400 °C. At 1650 °C, forsterite, alumina and carbon black were almost completely reacted, and $MgAl_2O4$, SiC and FeSi formed. Compared with MA1, carbon black, reducing the direct contact of forsterite and alumina, which influenced the formation of spinel, and acted as a reducer generated SiC and FeSi. The products were powder not bulk at 1400 °C and 1650 °C. It could been considered that the blocks were disrupted by amount of pores that carbothermal synthesis process produced.

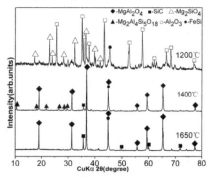

Fig.4 XRD patterns of sample MA2 heated under argon atmosphere at different temperature for 3hrs

Fig.5 shows SEM micrographs of sample MA2 at different temperature. The silicon carbide whiskers (Fig.5a) and strip SiC (Fig.5c) formed at 1400 °C and 1650 °C, respectivly. Scanning electron microscope observations show clearly that the tip of the whiskers is usually ellipsoidal (Fig. 5b). The silicon carbide whiskers ending with an FeSi droplet and the formation accorded to the vapour-liquid-solid (VLS) mechanism. [14]

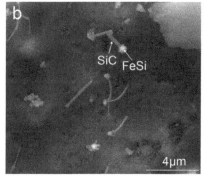

Fig. 5 SEM micrograph of sample MA2 under argon atmosphere
(a, b) 1400 °C, (c) 1650 °C

CONCLUSIONS

From above discussions, it can be concluded that MgAl2O4 crystallite formed by solid phase reaction, most like the reaction of the forsterite and alumina, its growth occurred by solid-liquid phase reaction. The silicon carbide ending with an FeSi droplet and the formation accorded to the vapour-liquid-solid (VLS) mechanism.

ACKNOWLEDGEMENT

This work was supported by the National Key Technology R&D Program (Grant No. 2011BAB03B04) and the National Science Foundation of China (NSFC) (Grant No. 50874084).

REFERENCES

[1]B.Y. Ma, Q. Zhu, Y. Sun etc, Synthesis of Al$_2$O$_3$-SiC Composite and Its Effect on the Properties of Low-carbon MgO-C Refractories, *J. Mater. Sci. Technol.*, 26, 715-720 (2010).

[2]T. Yamamura, T. Hamazaki, H. Kato, Alumina spinel castable refractories in steel teeming ladles, *Taikabutsu Overseas*, 12, 21–27 (1992).

[3]H. Sarpoolaky, S. Zhang, W.E. Lee, Corrosion of high alumina and near stoichiometric spinels in iron-containing silicate slags, *J. Eur. Ceram. Soc.*, 23, 293-300 (2003).

[4]C. Baudin, R.R. Martinez, P. Pena, High-temperature mechanical behavior of stoichiometric magnesium spinel, *J. Am. Ceram. Soc.*, 78, 1857-1862 (1995).

[5]P.Y. Lee, H. Suematsu, T. Yano and K. Yatsui, Synthesis and characterization of nanocrystalline MgAl$_2$O$_4$ spinel by polymerized complex method, *J. Nanopart. Res.*, 8, 911-917 (2006).

[6]Q.Y. Chen, C.M. Meng, T.C. Lu, X.H. Chang, G.F. Ji, L. Zhang, F. Zhao, Enhancement of sintering ability of magnesium aluminate spinel (MgAl$_2$O$_4$) ceramic nanopowders by shock compression[J]. *Powder technol.*, 200, 91-95 (2010).

[7]S. Zhang, N.J. Marriott, W.E. Lee, Thermochemistry and microstucture of MgO-C refractories containing various antioxidants, *J. Eur. Ceram. Soc.*, 21, 1037-1047 (2001).

[8]Yamaguchi A, Behaviors of SiC and Al added to carbon containing refractories, *Taikabutsu Overseas*, 4, 14-18 (1984).

[9]A.S. Gokce, C. Gurcan , S. Ozgen, S. Aydin, The effect of antioxidants on the oxidation behaviour ofmagnesia–carbon refractory bricks, *Ceram. Int.*, 34 (2008) 323–330.

[10]W.J. Yuan, P. Liang, C.J. Deng etc., Effects the atmosphere on synthesis of MgO-SiC-C refractory powders by using forsterite and carbon black, *Interceram Refractories Manual*, 56-58 (2011).

[11]Y.F. Wei, C.J. Deng H.X. Zhu etc., Synthesizing MgO-SiC-C refractory composite powders from forsterite and carbon black, *J. Xi'an Jiaotong Univ.*, 449, 105-109, (2010).

[12]P. Liang, M. Qiang, H.X. Zhu etc., Synthesis of MgAl$_2$O$_4$-SiC-C refractory by forsterite, bauxite and carbon black, *J. Mater. Eng.*, S2, 449-453 (2010).

[13]H.J. Duan, H.X. Zhu, P. Liang etc. Synthesis of MgAl$_2$O$_4$-SiC-C Refractory Powders by Using Forsterite, Alumina and Carbon Black. *Adv. Mater. Res.*, 399-401, 838-841(2012).

[14]T. Belmonte, L. Bonnetain, *J.L. Gieoux*, Synthesis of silicon carbide whiskers using the vapour-liquid-solid mechanism in a silicon-rich droplet, *J. Mater. Sci.*, 30, 2367-2371 (1996).

JOINING OF ALUMINA BY POLYCARBOSILANE AND SILOXANE INCLUDING PHENYL GROUPS

Ken'ichiro Kita and Naoki Kondo

National Institute of Advanced Industrial Science and Technology (AIST)

2266-98 Shimo-shidami, Moriyama-ku, Nagoya, 463-8560, JAPAN

ABSTRACT

This paper investigates the direct reaction between aluminum and silica (Al-SiO$_2$) or aluminum and silicon oxycarbide (AlSiOC) as a potential approach for joining alumina. A polymer blend containing polycarbosilane and polymethylphenylsiloxane was suitable for the surface modification of alumina to yield silicon oxycarbide. In the case of a reaction between Al and SiO$_2$, the joining layer was transformed to uniform alumina silicate and the average bending strength was ~177 MPa. In the case of Al-SiOC, a transformation into two different types of joining layer could be achieved. One was a metal-silicon layer whose average bending strength was ~252 MPa, while the other was a carbon-substance layer whose average bending strength was ~58 MPa. The metal-silicon layer was obtained by the completion of the reduction reaction between the metallic Al and the SiOC layer, and the carbon-substance layer was obtained as a by-product of the abovementioned reduction reaction.

1. INTRODUCTION

Ceramics include many desirable properties such as good thermal stability, high hardness, abrasive resistance, and lightweight in comparison with metallic materials. Therefore, ceramic materials may substitute heavy-metal components in various production chains such as their use in different stages of semiconductor processing, thereby leading to energy savings and reduction of environmental pollution [1]. However, large ceramic products are difficult to manufacture, because an electric furnace larger than the product must be prepared, which, however, leads to high cost and energy consumption. Moreover, this method is ineffective and wasteful.

To manufacture large ceramic products, a number of ceramic joining methods have been reported [2–5]. Among them, ceramic joining using a polymer precursor is attractive because of its superior features. For example, this method allows skipping grinding before joining because a polymer can intrude into tiny cracks at the ceramic surface. Moreover, it can be applied to both very small and very large ceramic components that are difficult to uniformly grind. Besides, polymers contain atoms that are easy to diffuse into ceramic materials, thereby enabling a tailored modification of the ceramic surface [6]. All these advantages increase the applicability of this method. However, only a few investigations on ceramic joining using a precursor polymer have been reported [7–9]. Moreover, the average strength of the joining area was <150 MPa because of the tiny cracks that developed during heating in the joining layer [10]. Therefore, a novel method for ceramic joining is required.

To achieve this purpose, we diverted our attention to a reaction between a light metal and a ceramic, as illustrated in the following chemical formula:

$$SiO_2(s) + 4/3\ Al(l) \rightarrow Si(s) + 2/3\ Al_2O_3(s) \qquad \Delta_f G° = -200.3\ kJ/mol \qquad (1)$$

This chemical equation shows a direct reaction between silica and aluminum metal without gas generation [11]. The tiny cracks in the joined ceramics using a precursor polymer were caused by gas generation from polymer during heating. Therefore, the joining of alumina can be carried out without cracks by modifying the alumina surface to silica beforehand. A polymer precursor enables to modify the alumina surface to silicon carbide, so that a further change of the surface to silica would be easy to carry out [12].

Therefore, in this study, we investigated the joining of ceramics by the direct reaction of aluminum foil and a modified membrane derived from a polymer precursor according to Eq. (1) to achieve high-strength joining of alumina.

2. EXPERIMENTAL PROCEDURE

Figure 1 shows the experimental procedure of this study. To effectively carry out the direct reaction, the modified surface of alumina must contain large amount of oxygen in various forms such as silica. Therefore, an additional oxidation curing by heating should be carried out after the surface modification solely by using polycarbosilane (PCS) because it can be transformed into silicon carbide [13]. We prepared a polymer blend containing PCS and polymethylphenylsiloxane (PMPhS) for surface modification. PMPhS is a type of siloxane that contains a large amount of oxygen. Moreover, the solubility of PMPhS in PCS is very high [14]. We produced a polymer blend with 70wt% of PCS and 30wt% of PMPhS, referred to as PS30. By using a polymer blend including siloxane, such as PS30, silicon oxycarbide can be easily obtained, compared to using only PCS [15]. A surface modification by solely using siloxane failed and exfoliation between siloxane and alumina was easily carried out. PCS is essential to successfully carry out surface modifications of alumina [12].

Bulk alumina whose purity was >99.9% was cut into alumina pieces (length: 20 mm; width: 30 mm; height: 20 mm). A surface with 600 mm² area was abraded with a grinder and the following investigations were carried out on this abraded surface. Bulk alumina specimens were dipped in PCS (NIPUSI-Type A, Nippon Carbon, Japan) or PS30 containing PCS and commercial PMPhS (KF-54, Shin-Etsu Chemicals Co. Ltd., Japan). After drying, the samples were cured in an air flow. The heating rate of the samples during curing was fixed at 8 K/min up to 473 K; subsequently, the temperature was maintained constant for 1 h. After curing, these samples were pyrolyzed at 1273 K for 1 h under Ar atmosphere and repyrolyzed at 1473 K for 2 h in an air flow. The ceramization of these polymers was completed by the first pyrolysis while the oxidation of the ceramic membranes formed from these polymers was completed by the repyrolysis.

An aluminum foil with a thickness of ~11 μm was sandwiched between the ceramic membranes on the top of the abraded surfaces of the alumina samples and heated to 1073 K for 2 h in vacuum for joining. A previous report shows that the chemical reaction according to Eq. (1) is effectively progressed by heating to >1373 K [11]. In order to prevent metal-silicon from effectively transforming to alumina, the joining temperature was set <1373 K. Moreover, aluminum along with the modified surfaces can be converted to various thin ceramic layers such as alumina and mullite [16]. Hereafter, the joined samples using the PCS and PS30 polymers are referred to as "PCS sample" and "PS30 sample", respectively.

After the above-mentioned process, the joining area of the pyrolyzed surface was observed by scanning electronic microscopy (SEM; JEM-5600, JEOL, Japan), energy dispersive X-ray spectroscopy (EDS; JEM-2300, JEOL, Japan), and X-ray diffraction (XRD; RINT2500, Rigaku Corporation, Japan). Finally, a 4-point bending test was performed on a testing machine model JIS R 1601:2008.

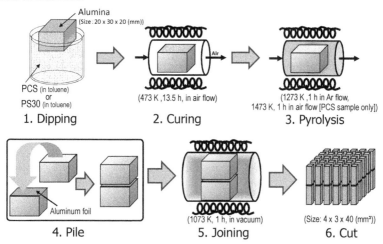

Figure 1 Outline of the experiment carried out in the present paper.

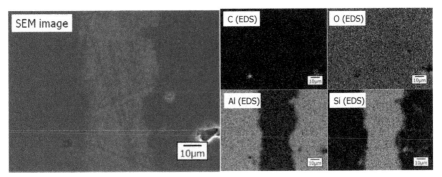

Figure 2 Cross-sectional SEM image and the corresponding EDS mappings of the joining area of PCS sample.

3. RESULTS AND DISCUSSION

To understand the composition, the joining areas of the samples were observed. Figure 2 shows a cross-sectional SEM image (left-hand side) and the corresponding EDS mappings (right-hand side) of the joining area of the PCS sample. The joining area could be recognized by a lighter color than that of aluminum, and the thickness of the area was ~20 μm. This SEM image (Fig. 2) reveals that no cracks occurred at the interface between the joining area and alumina. The EDS mappings of aluminum, silicon, carbon, and oxygen in the joining area were observed (right-hand side of the SEM image in Fig. 2). The spectrum of oxygen shows a uniform distribution in the alumina and the joining area. In the case of aluminum, only a little peak was remained in the joining area in spite of the presence of an aluminum foil sandwiched between the modified alumina surfaces before joining. In the EDS mapping of silicon, the peak intensity mainly existed within the joining area. This result is expected because silicon atoms were derived from the modified surfaces only. However, additional peaks could be observed originating from the alumina area. These results reveal that SiO_2 of the modified surfaces and aluminum metal were integrated and uniformly diffused in the joining area.

In the case of PS30, two types of samples were observed. Figure 3 shows the cross-sectional SEM image (left-hand side) and the corresponding EDS mappings (right-hand side) of the joining area of a type of PS30 samples. The width of the area was ~30 μm. The interlayer between the joining and the alumina area had no crack. However, some features of this sample were different than the PCS sample. For example, neither a peak of oxygen nor of aluminum was observed in the joining layer. Moreover, only a strong silicon peak was observed. It appears that the joining area was occupied by silicon only, and therefore, the reduction reaction described by Eq. (1) was possibly completed in spite of the relatively low temperature of 1073 K during joining. Hereafter, this type of samples is referred to as "PS30-A sample."

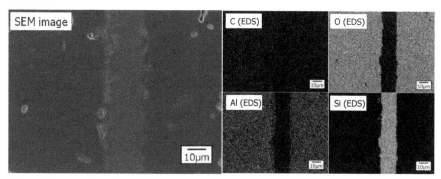

Figure 3 Cross-sectional SEM image and the corresponding EDS mappings of the joining area of PS30-A sample.

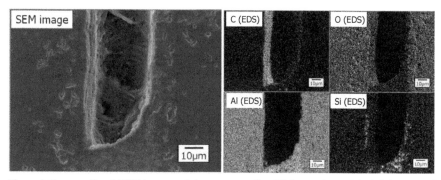

Figure 4 Cross-sectional SEM image and the corresponding EDS mappings of the joining area of PS30-B sample.

Figure 4 shows the cross-sectional SEM image (left-hand side) and corresponding EDS mappings (right-hand side) of the joining area of the other type of PS30 samples. This sample also included some differences with respect to both the PCS and PS30-A samples. The joining area sunk in and transformed to a ditch. Besides, a detailed analysis of the SEM image showed that the bottom of the ditch had a porous structure. In addition, a strong spectrum of carbon could be observed at the periphery of the ditch whereas strong peaks of alumina and silicon, as found in the PCS and the PS30-A sample, could not be observed in this region of high carbon concentration. This result reveals that the joining area of these types of samples comprised a carbon component and a porous structure. Hereafter, such samples are referred to as "PS30-B sample."

Figure 5 shows the XRD patterns of the joining area of PCS, PS30-A, and PS30-B samples. The chemical composition of the joining layers could be identified by XRD. All the XRD patterns show alumina and aluminum metal as background. In the patterns of the PCS and PS30-A samples, the occurrence of two broad and smeared out peaks around $\theta = 12.0°$ and $21.0°$ reflect the presence of an amorphous compound [17–19]. Considering the results of the SEM images and EDS mappings, the compound was considered to be derived from amorphous aluminum silicate in the PCS sample and from the metal-silicon interface in the PS30-A sample. Moreover, the peaks of aluminum silicate are mainly located around 9.7°, 20°, 31.8°, 41.6°, and 51.2° in the PCS sample, as shown in Figure 5. Therefore, taking into account the existence of an amorphous phase, it can be concluded that the joining area of the PCS sample mainly consists of aluminum silicate. In the case of the PS30-A sample, the intensity of the peaks of silicon were remarkably increased when compared to those of the PCS sample. Besides, the joining areas consisting of dense silicon appear to be covered by another layer on both the sides [20,21], in agreement with the results of the SEM image and EDS mappings. However, in the case of the PS30-B sample, the diffraction intensities corresponding to the amorphous phase could not be recognized. Instead, an increase in the intensity of the carbon peaks could be observed. The peaks attributed to carbon have an angular position similar to those of aluminum oxycarbide, mainly appearing below 15° and no remarkable peaks except carbon in the pattern of PS30-B sample were observed [22,23].

The results of the 4-point bending tests are shown in Figure 6. In this graph, the standard deviations are shown as error bars. Besides, the alumina sample joined by PCS only is also shown in this figure for comparison [10]. The number of samples for this test was fixed to 12. The average strength of the PCS samples was ~177 MPa and the maximum strength was 239 MPa. This result shows that these samples have higher strength than those obtained from the alumina samples joined by using polymer and heating only. The increase in the strength is in line with our expectations. Besides, most of the samples broke within the joining layer; therefore, there is a possibility that the actual strength was higher than the values determined above. In the case of the PS30-A sample, the average strength was only 252 MPa while the maximum strength was 395 MPa. Most of these samples broke away from alumina. Therefore, it was considered that the strength of this joining layer was higher than alumina. This strength seemed to be derived from the metal-Si interface, the main constituent of the joining layer. However, in the case of PS30-B, the average strength resulting from the bending test was only 58 MPa. The site of fracture in all these samples was the joining area. This result is acceptable because the joining layer consisted of carbon components and a porous structure forming a ditch.

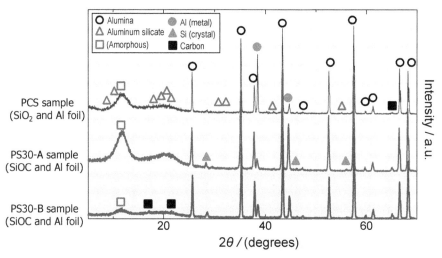

Figure 5 XRD patterns of the joining area of PCS and PS30 samples.

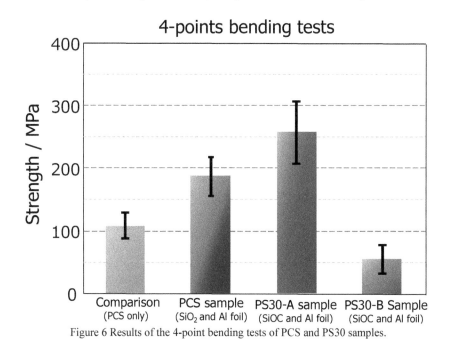

Figure 6 Results of the 4-point bending tests of PCS and PS30 samples.

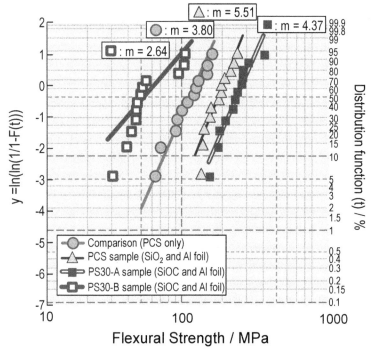

Figure 7 Weibull distribution of the 4-point bending tests of PCS and PS30 samples.

Figure 7 shows the Weibull distribution of the 4-point bending tests performed on the PCS and PS30 samples. The alumina sample joined by PCS only is also shown in this figure for comparison. These plots were calculated by median rank method commonly used in similar studies [24]. The values of "m" shown in the individual plots of Fig. 7 are the Weibull modulus. The Weibull modulus of the PCS sample was 5.51 and that of the PS30-A sample 4.37. These values are higher than the reference value obtained from the sample of joined alumina derived from polymer dipping and heating to 1873 K. However, the Weibull modulus of PS30-B was 2.64, lower than the reference value of 3.80. These results show that joining by the direct reaction between the aluminum foil and the surfaces modified to silica, such as in the case of the PCS sample, is better for reducing unevenness in the bending strength of the joined materials. However, in the case of the direct reaction between the aluminum foil and the surfaces modified to silicon oxycarbide, such as in the case of the PS30 samples, the joining area that can decrease the unevenness intermingled with the one that can increase the unevenness. Therefore, a novel method that enables the removal of carbon components and prevents the formation of a ditch with a porous structure, as observed in the joining area of the PS30-B sample, is required.

To find a clue to such a novel method, we considered the reaction between the aluminum foil and silicon oxycarbide during heating. The clear difference between silica and silicon oxycarbide is because of the presence of carbon in the latter. Carbon heated under an inert atmosphere can be chemically active and may force some oxides to be deoxidized. This deoxidization by carbon is commonly used for the removal of the strong oxide layer at the surface of aluminum metal and is responsible for the increased wettability between alumina and aluminum [25,26]. Therefore, it can be considered that the carbon of silicon oxycarbide reacts and deoxidizes the oxides at the surface of the aluminum foil.

Aluminum foil could directly come in touch with silicon oxycarbide because of deoxidization, and aluminum reacted with silicon oxycarbide to form a metal-silicon interface, alumina, and aluminum carbide. The PS30-A samples showed the existence of a metal-silicon interface; therefore, it was generally believed that the by-products derived from this reaction were alumina and aluminum carbide. Although a previous study shows that the effective reduction of silica by using aluminum needs heating to a temperature >1373 K, the reduction may progress because this temperature is needed for the reduction of the oxide on the top of an aluminum surface [11]. Moreover, although the generation of aluminum carbide is not favored from the viewpoint of thermodynamics, the rapid diffusion of aluminum atoms because of the deoxidization of the oxides enables the generation of aluminum carbide from aluminum and excess carbon [27]. Amorphous SiC pyrolyzed at 1073 K, derived from pure PCS, contains excess carbon (including hydrocarbon); therefore, silicon oxycarbide derived from PS30 seems to have excess carbon because of the phenyl groups [28].

It is accepted that the amount of carbon in a joining layer, such as in the case of the PS30-B sample, was derived from the excess carbon in the modified surface of alumina. If the amount of carbon in the modified surface can be adjusted to a level so that only deoxidization of the oxide at the aluminum surface takes place, joined alumina with a joining layer consisting only of a metal-silicon interface can be obtained.

4. CONCLUSION

Based on the abovementioned results, the direct reaction between the aluminum foil and the modified alumina surfaces could be carried out and applied as a novel ceramic joining technique.

The silica membrane and the aluminum foil could be transformed to uniformly distributed aluminum silicate in which the average strength of the joining layer was 177 MPa. On the other hand, a silicon oxycarbide surface on alumina and the aluminum foil can be transformed to a metal-silicon-carbon substance with pores. A metal-silicon interface was derived from the reduction reaction of aluminum and silicon oxycarbide, and aluminum carbide was generated as a byproduct. Therefore, two types of joining layers could be obtained by this reaction. The average strength of the metal-silicon joining area was 252 MPa, stronger than alumina; however, the average strength of the carbide joining area amounted to only 58 MPa.

ACKNOWLEDGEMENT

This research was supported by METI and NEDO, Japan, as part of the Project for the Development of Innovative Ceramics Manufacturing Technologies for Energy Saving.

References

[1] N. Tikul, and P. Srichandr, "Assessing the environmental impact of ceramic tile production in Thailand" *J. Ceram. Soc. Jpn.*, **118**, 887-894 (2010).

[2] M. Nicolas, "The strength of metal/alumina interfaces", *J. Mater. Sci.*, **3** (1968), 571-576

[3] L. S. D. Glasser, J. A. Gard, and E. E. Lachowski, "The reaction of zinc oxide and zinc dust with sodium silicate solution" *J. Appl. Chem. Biotechnol.*, **28** (1978), 799-810.

[4] J. T. Klomp and T. P. J. Botden, "Sealing Pure Alumina Ceramics to Metals", *Ceram. Bull.*, **49** (1970), 204-211.

[5] M. Naka, Y. Hirono, and I. Okamoto, "Joining of Alumina/Alumina Using Al-Cu Filler Metal and Its Application to Joining of Alumina/Aluminum", *J. Jpn. Weld. Soc.*, **5** (1987), 93-98.

[6] K. Kita, N. Kondo, Y. Izutsu, H. Kita, "Investigation of the properties of SiC membrane on alumina by using polycarbosilane", *Mater. Lett.*, **75** (2012), 134-136.

[7] S. Yajima, K. Okamura, T. Shishido, Y. Hasegawa, and T. Matsuzawa, "Joining of SiC to SiC using polybolosiloxane", *Am. Ceram. Soc. Bull.*, **60** (1981), 253.

[8] E. Anderson, S. Ijadi-Maghsoodi, O. Ünai, M. Nostrati, and W. E. Bustamante, "Ceramic Joining," *Ceramic Transactions Vol.77*, ed. I. E. Reimanis, C. H. Henager and A. P. Tomsia (Westville, OH: The American Ceramics Society, 1997), 25-40

[9] P. Colombo, V. Sglavo, E. Pippel, and A. Donato, "Joining of reaction-bonded silicon carbide using a preceramic polymer", *J. Mater. Sci.*, **33** (1998), 2405-2412.

[10] K. Kita, N. Kondo, Y. Izutsu, H. Kita, "Joining of alumina by using organometallic polymer", *J. Ceram. Soc. Jpn.*, **119** (2011), 658-66.

[11] T. Okutani, "Utilization of Silica in Rice Hulls as Raw Materials for Silicon Semiconductors", *J. Metal. Mater. Mineral*, **19** (2009), 51-59.

[12] K. Kita, N. Kondo, H. Hyuga, Y. Izutsu, H. Kita, "Study of modification on alumina surface by using of organosilicon polymer", *J. Ceram. Soc. Jpn.*, **119** (2011) , 378-381.

[13] S. Yajima, J. Hayashi, M. Omori and K. Okamura, "Development of a silicon carbide fibre with high tensile strength," *Nature,* **261** (1976), 683-685.

[14] K. Kita, M. Narisawa, A. Nakahira, H. Mabuchi, M. Sugimoto, and M. Yoshikawa, "Synthesis and properties of ceramic fibers from polycarbosilane/ polymethylphenylsiloxane polymer blends", *J. Mater. Sci.*, **45** (2010), 3397-3404.

[15] K. Kita, M. Narisawa, A. Nakahira, H. Mabuchi, M. Itoh, M. Sugimoto, M. Yoshikawa, "High-temperature pyrolysis of ceramic fibers derived from polycarbosilane– polymethylhydrosiloxane polymer blends with porous structures", *J. Mater. Sci.*, **45** (2010), 139-145.

[16] S. Maitra, R. Shibayan, and A. K. Bandyapadhyay, "Synthesis of mullite from calcined alumina, silica and aluminium powder", *Ind. Ceram.*, **24** (2004), 39-42

[17] R. J. P. Correiu, P. Gerbier, C. Guerin, and B. Hammer, "From Preceramic Polymers with Interpenetrating Networks to SiC/MC Nanocomposites", *Chem. Mater.*, **12** (2000), 805-811

[18] M. Narisawa, R. Sumimoto, K. Kita, H. Kado, H. Mabuchi, and Y.-W. Kim, "Melt Spinning and Metal Chloride Vapor Curing Process on Polymethylsilsesquioxane as Si-O-C Fiber Precursor", *J. Appli. Polym. Sci.*, **114** (2009), 2600-2607

[19] X. Yuan, S. Chen, X. Zhang, and T. Jin, "Joining SiC ceramics with silicon resin YR3184", *Ceram. Int.*, **35** (2009), 3241-3245

[20] K. Laaziri, S. Kycia, S. Roorda, M. Chicoine, J. L. Robertson, J. Wang, and S. C. Moss, "High-energy x-ray diffraction study of pure amorphous silicon", *Phys. Rev. B*, **60** (1999), 520-533

[21] A. Vivas Hernandez, T. V. Torchynska, A. L. Quintos Vazquez, Y. Matsumoto 2, L. Khomenkova, and L. Shcherbina, "Emission and structure investigations of Si nano-crystals embedded in amorphous silicon", *J. Phys. Conf. Ser.*, **61** (2007), 1231-1235

[22] C. Ji, Y. Ma, M. –C. Chyu, R. Knundson, and H. Zhu, "X-ray diffraction study of aluminum carbide powder to 50 GPa", *J. Appl. Phys.*, **106** (2009), 083511.

[23] J. H. Cox, and L. M. Pidgeon, "THE X-RAY DIFFRACTION PATTERND OF ALUMINUM CARBIDE Al₄C₃ AND ALUMINUM OXYCARBIDE AL₄O₄C", *Can. J. Chem.*, **41** (1963), 1414-1416.

[24] K. M. Entwistle, "The fracture stress of float glass" *J. Mater. Sci.*, **28** (1993), 2007-2012

[25] J. J. Brennan, and J. A. Pask, "Effect of Nature of Surfaces on Wetting of Sapphire by Liquid Aluminum", *J. Am. Ceram. Soc.*, **51** (1968), 569-573

[26] M. G. Nicholas, D. A. Mortimer, L. M. Jones, and R. M. Crispin, "Some observation on the wetting and bonding of nitride ceramic", *J. Mater. Sci.*, **25** (1990), 2679-2689

[27] K. C. H. Kumar, and V. Raghavan, "Thermodynamic Analysis of the Al-C-Fe System", *J. Phase Equilibria*, **12** (1991), 275-286

[28] E. Bouillon, F. Langlais, R. Pailler, R. Naslain, F. Cruege, P. V. Huong, J. C. Sarthou, A. Delpuech, C. Laffon, P. Lagarge, M. Monthioux, and A. Oberlin, "Conversion mechanisms of a polycarbosilane precursor into an SiC-based ceramic material", *J. Mater. Sci.*, **26** (1991), 1333-1345

MICROWAVE JOINING OF ALUMINA USING A LIQUID PHASE SINTERED ALUMINA INSERT

Naoki KONDO, Mikinori HOTTA, Hideki HYUGA, Kiyoshi HIRAO and Hideki KITA
National Institute of Advanced Industrial Science and Technology (AIST)
Shimo-shidami 2266-98, Moriyama-ku, Nagoya 463-8560, Japan

ABSTRACT

High-purity alumina (>99.5%) was joined using liquid-phase-sintered alumina as an insert material. Dense liquid phase–sintered alumina, containing SiO_2 and MgO as sintering additives, was placed between the bulk alumina bodies to make a joint. These were placed in a microwave furnace and joining was performed at 1650°C for 30 min under a uniaxial mechanical pressure of 0.17 MPa. The joining was successful and no cracks or separations were found between the bulk and the joint. Grain growth in the bulk alumina, pore formation in the joint, and diffusion of Si and Mg from joint to bulk were examined. The strength of the joined alumina was 200 MPa; therefore, high strength was achieved.

INTRODUCTION

Joining technology is critically important when fabricating ceramic components. Alumina is an important structural ceramic; therefore, a large number of joining techniques have been developed.[1, 2] Joining techniques must be based on usage; for example, alumina used at high temperatures must have a heat resistant joint. Liquid-phase-sintered alumina insert and alumina-zirconia composite insert[3] are the most promising candidates for joining alumina for use at high temperatures.

Microwave heating is an interesting heating technique, wherein a material can be heated locally and rapidly. There are some reports of microwave heating being used to join alumina.[4, 5] A single-mode microwave that can locally heat the alumina joint was used for joining in these cases. However, the heating of large areas was difficult. On the other hand, a multi-mode microwave can heat larger areas using a susceptor. This technique was successfully applied to join silicon nitride with a SiAlON glass joint[6] using silicon carbide as the susceptor. Silicon carbide was placed around the joint. Silicon nitride did not absorb microwaves, but silicon carbide absorbed microwaves and were heated, thereby achieving local heating. Joining of alumina is also expected to proceed in a similar manner.

In this study, high-purity alumina was joined with a liquid-phase-sintered alumina insert using microwave heating. The microstructure of the joined alumina was investigated and its strength was examined.

EXPERIMENTAL

The bulk alumina bodies used in this study were fabricated from high-purity (>99.5%) alumina powder, AL160SG4, by Showa Denko K.K. Dense alumina-sintered bodies of 30 × 10 × 20 mm were prepared by sintering the powder at 1600°C for 2 h. The surfaces (30 × 10 mm) used for joining were machined with a #200 whetstone.

A liquid-phase-sintered alumina plate insert was fabricated from the mixed powder, alumina – 5 wt.% Spinel ($MgAl_2O_4$) – 4 wt.% SiO_2. A 30 × 10 × 1 mm plate was prepared by sintering the powder at 1600°C for 2 h. The surfaces (30 × 10 mm) used for joining were machined with a #200 whetstone.

Alumina/zirconia composite plates were used as susceptors and alumina fiberboard was used as the insulator. Alumina, liquid-phase-sintered alumina, and alumina fiberboard are poor microwave absorbers at low temperatures. On the other hand, zirconia grains in the alumina/zirconia susceptor are good microwave absorbers and were heated by microwave radiation.

Liquid-phase-sintered alumina was placed between two alumina bodies to make a joint. The joint plane was horizontally aligned. Four side (vertical) planes of the blocks were surrounded by the susceptor plates. This configuration was placed inside the alumina fiberboard and placed in a microwave furnace. A uniaxial low mechanical pressure of 0.17 MPa was applied to the joint plane by placing an alumina rod, steel rod, and steel weight on the blocks.

A microwave furnace with four magnetron sources (frequency: 2.45 GHz, maximum output: 1.5 kW × 4 (6.0 kW in total)) was used for heating. The temperature inside the susceptor plates (without alumina bodies) was measured using thermocouples up to 1500°C, and that outside the plate was measured using a pyrometer. The temperature was corrected by measuring the temperature difference between the thermocouples and pyrometer. A temperature higher than 1500°C was assumed from the result.

The configuration of alumina bodies was heated up to 1650°C in 1 h by controlling the microwave output to avoid thermal shock fractures. It was then soaked at 1650°°C for 30 min.

Specimens for microstructure observations and strength measurement were cut from the joined block. Microstructure observations were performed by optical microscopy (OM) and scanning electron microscopy (SEM). The composition was examined by energy dispersive x-ray spectroscopy (EDX). Specimens of $3 \times 4 \times 40$ mm were prepared for strength measurement. These specimens had a joint at the center of a bend bar. The four-point bending strength was measured in accordance with JIS R1601 using outer and inner spans of 30 and 10 mm, respectively, and at a displacement rate of 0.5 mm min^{-1}.

RESULTS AND DISCUSSION

The alumina blocks appeared to be well-joined after the joining procedure. Bend test specimens were cut from the joined blocks for strength measurement. The post-bend test specimens are shown in **Fig. 1**. Joints from the liquid-phase-sintered alumina insert are shown as white lines in the center of the specimens. Both sides of each specimen were composed of creamy yellow high-purity bulk alumina. The bulk alumina near the joint turned white after joining. The color change region was nearly 5 mm from the joint interface.

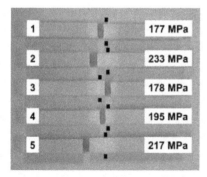

Fig. 1
Joined alumina specimens after bending strength measurement. Insert positions are indicated by black square marks. The width of the marks was the same as that of the insert.

The microstructure of the joined alumina was examined by OM (**Fig. 2**) and SEM (**Fig. 3**). The specimen was prepared by polishing followed by thermal etching. The joint interface between the joint and bulk was easily found because the grain size in the joint was larger than that in the bulk, and many pores existed in the joint. No crack or separation was found at the joint interface between the bulk and the joint. The grain size of bulk alumina near the joint (~0.3 mm away from the joint interface) was larger than that far from the joint (~5 mm away). The grain size at 5 mm away from the joint was the same as that farther in the bulk. In addition, grain boundaries near the joint were clearly observed. Pore distribution in the joint was uniform. The liquid-phase-sintered alumina insert used for joining was almost dense; therefore, the pores seemed to have formed during the joining procedure.

The composition change across the joint interface was examined by SEM-EDX. As shown in **Fig. 3**, counts of Al, O, Mg, and Si were slightly higher in the bulk than in the joint. The existence of pores in the joint may reduce the counts. Mg and Si were only located in the insert before joining, not in the bulk. Therefore, Mg and Si seemed to have diffused into the bulk from the insert during the joining procedure.

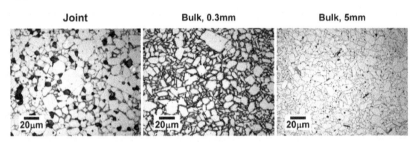

Fig. 2   Microstructure of the joined body.
The "Joint" micrograph was taken away from the center of the joint.
The "Bulk" micrographs were taken 0.3 and 5 mm away from the joint interface in the bulk, respectively.

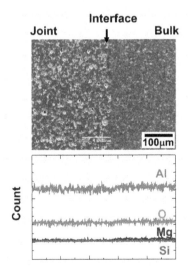

Fig.3
(Upper) SEM micrograph around the joint interface of the joined alumina.
(Lower) Counts of Al, O, Mg, and Si by EDX line analysis.

As shown in **Fig.3**, Mg and Si diffused into the bulk. Their diffusion depths were examined by EDX. The EDX intensities from several points are shown in **Fig. 4**. Mg was clearly detected at the 0.5 and 1 mm positions, but was not detected at 2 mm from the joint interface. The intensity of Si showed a similar characteristic to Mg, though it was weaker than Mg. Therefore, the diffusion depths of Mg and Si from the insert to the bulk were at least 1 mm. The color change region of bulk alumina, shown in **Fig.1**, expanded to about 5 mm from the joint interface. The grain size 5 mm away from the joint, shown in **Fig. 2**, was the same as that farther away in the bulk. Therefore, the diffusion depth should be almost 5 mm. The sensitivity of EDX was too low to detect Mg and Si at positions further than 2 mm.

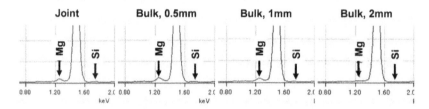

Fig. 4   EDX intensities of the joined body.
The intensities were taken from a rectangular area of 500×370 μm.
The "Joint" intensity was taken from the center of the joint.
The "Bulk" intensities were taken 0.5, 1, and 2 mm away from the joint interface in the bulk.

As mentioned above, two characteristic microstructures were observed: grain growth near the joint and pores in the joint. Grain growth near the joint seemed to be enhanced by the existence of a liquid phase, which was diffused into the bulk from the insert. Grooving of grain boundaries during thermal etching was also enhanced by the liquid phase. This led to clear observation of the grain boundary in **Fig.1**. Two possible reasons for the insert becoming porous were considered. One is the diffusion of the liquid phase from the insert to the bulk. The triple points of the grains where the liquid phase existed before joining changed to pores by diffusing out of the liquid phase after the joining procedure. The other is over-sintering of liquid-phase-sintered alumina. Grain growth occurred during the joining procedure, resulting in over-sintering. Additionally, grain growth might be enhanced by microwave heating because liquid phases generally absorb microwaves well and heat up, compared to the solid phase.[5] This grain growth led to the formation of pores at the triple points of grains.

The average strength of the joined specimen was 200 MPa. The strengths of the individual specimens are indicated in **Fig. 1**. Two specimens, Nos. 1 and 4, were fractured from the joint interface. The other three were fractured from the bulk. The lowest strengths fractured from the interface and the bulk were almost identical. Therefore, the joint interface was strong enough and good joining was achieved.

CONCLUSIONS
The joining of high-purity alumina (>99.5%) using liquid-phase-sintered alumina, containing SiO_2 and MgO as sintering additives, as an insert was successfully performed. It was possible to join alumina using a combination of temperature (1650°C), time (30 min), pressure (0.17MPa), and heating method (multimode microwave heating with a susceptor). No cracks or separations were found between the bulk and the joint. Si and Mg diffused from the insert into the bulk, enhancing

grain growth in the bulk near the joint. Pores were formed in the joint by diffusion of the liquid phase or over-sintering of alumina. The strength of the joined alumina was 200 MPa; therefore, high strength was achieved.

REFERENCES
[1] Akselsen OM, "Review: Diffusion bonding of ceramics," *J. Mater. Sci.*, **27**, 569-579, (1992).
[2] Loehman RE, "Recent progress in ceramic joining," *Key Eng. Mater.*, **161-163**, 657-661, (1999).
[3] Kondo N, Hotta M, Hyuga H, Hirao K, Kita H, in preparation.
[4] Fukushima H, Yamanaka T, Matsui M, "Microwave heating of ceramics and its application to joining," *J. Mater. Res.*, **5**, 397-405, (1990)
[5] Binner JGP, Fernie JA, Whitaker PA, Cross TE, "The effect of composition on the microwave bonding of alumina ceramics," *J. Mater. Sci.*, **33**, 3017-3029, (1998).
[6] Kondo N, Hyuga H, Kita H, Hirao K, "Joining of silicon nitride by microwave local heating," *J. Ceram. Soc. Japan.*, **118**, 959-962, (2010).

ACKNOWLEDGEMENT
This research was supported by METI and NEDO, Japan, as part of the Project, "Innovative Development of Ceramics Production Technology for Energy Saving."

JOINING OF SILICON NITRIDE LONG PIPES WITHOUT INSERT MATERIAL BY LOCAL HEATING TECHNIQUE

Mikinori HOTTA, Naoki KONDO, Hideki KITA, and Tatsuki OHJI
Advanced Manufacturing Research Institute, National Institute of Advanced Industrial Science and Technology (AIST)
2266-98 Shimo-Shidami, Moriyama-ku, Nagoya 463-8560, Japan

ABSTRACT
Silicon nitride pipes of 1 m in length were joined without using insert material to produce the long silicon nitride pipe by a local-heating joining technique. Commercially available silicon nitride pipes sintered with Y_2O_3 and Al_2O_3 additives were used for parent material. The silicon nitride pipes had rough or uneven end faces. Joining was performed by locally heating the joint region at 1600°C for 1 h with a mechanical pressure of 5 MPa in flowing N_2 gas. The microstructure of the joint region was observed using a scanning electron microscopy and the joining strength was measured by a four-point bending test. Voids were observed in the whole joint region of the joined silicon nitride pipe. When no voids were observed in the joint region, the border with the parent region was not clearly distinguishable. Flexural strength of the samples cut from the joined silicon nitride pipe indicated the average value of about 170 MPa. On six tested samples, all samples fractured from the joint region. The low joining strength and the fracture from the joint region are due to the existence of voids in the whole joint region. These results found that the successful joining of silicon nitride with rough or uneven joined faces requires the use of insert material.

INTRODUCTION

Silicon nitride ceramics are expected to be applied as industrial components for saving the energy for production and improving the quality of products in manufacturing industries, because of their excellent heat resistance, corrosion resistance, good wear resistance, high specific elastic modulus, and lightweight[1]. These ceramic components are often required to have large scale of more than 10 m in length. Joining is one of the most important technologies to obtain large-scaled ceramic components, because it is difficult to produce them as single units[2]. When producing large-scaled ceramic components, however, it is necessary to join ceramic units by locally heating the joint region in order to reduce the energy consumption and cost in the production process. In addition, the ceramic components are often exposed to a high-temperature and/or corrosive environments in manufacturing industries. Therefore, the joint layers require having sufficient thermal, mechanical and chemical properties which are similar to those of the parent ceramic units. Ideally, this can be achieved by forming the microstructure and chemical composition identical to each other.

There are many reports on joining of silicon nitride ceramics by heat-treatment. In these studies, however, high joining temperatures, high mechanical pressure, and/or long post heat-treatments have been required for obtaining high joining strength of silicon nitride. Furthermore, the joined samples of these studies were small and the whole of the samples were heated in conventional furnaces.

The authors recently reported on a local-heating joining technique using electric furnace equipment specially developed for joining long ceramic pipes[3,4]. Powder slurry of Si_3N_4-Y_2O_3-Al_2O_3-SiO_2 system was brush-coated on the rough or uneven end faces of the silicon nitride pipes as insert material. The advantage of this technique was easily joining ceramic pipes even with rough or uneven end faces for producing long silicon nitride tubular components. The joined 3 m long silicon nitride pipe was successfully produced without voids or cracks in the joint region, and the gap between the pipes with rough or uneven joined faces was filled by the slurry brush-coated as insert material. The samples cut from the silicon nitride pipes joined at 1600 and 1650°C showed flexural strength of about 680 MPa, which was similar to that of the parent silicon nitride bulk.

This paper studied the joining performance of silicon nitride pipes without using insert material by the local-heating joining technique. The silicon nitride pipes with rough or uneven end faces were joined by locally heating the joint region at a joining temperature of 1600°C, and microstructure of the joint region and joining strength of the joined sample were investigated.

EXPERIMENTAL PROCEDURE

Commercially available silicon nitride ceramic pipes sintered with Y_2O_3 and Al_2O_3 (SN-1 grade, Mitsui Mining & Smelting Co., Ltd., Tokyo, Japan) were used as the joined pipes. The pipes were 1 m in length, and had outer and inner diameters of 28 and 18 mm, respectively. The silicon nitride pipes had rough or uneven joint surface. The two pipes were fixed to the local heating equipment, and the joint region of the pipes was inserted into the electric furnace in the equipment. The silicon nitride pipes were joined at temperature of 1600°C for a holding time of 1 h in flowing N_2 gas at a rate of 5 L/min. A mechanical pressure of 5 MPa was applied to the joint surface during the joining. The joined silicon nitride pipe was cut perpendicular to the joint surface to observe the microstructure of the joint region and measure the flexural strength of the joined pipe. The surface of the samples was polished with diamond slurry and then etched by plasma in CF_4 gas. The etched surface of the samples was observed by scanning electron microscopy (SEM). The flexural strength of the samples was measured at room temperature using a four-point bending method on six samples; the tested samples had dimensions of 3 mm x 4 mm x 40 mm, and the joint region was at the center of the bending bar.

RESULTS AND DISCUSSION

Figure 1 shows SEM micrographs of the joint region of the sample joined without using insert material at 1600°C. Voids were clearly observed in the whole joint region (arrows in Figs. 1 (a) and (b)). The thickness of the voids was several microns. The formation of the voids is mainly attributed to the roughness or unevenness of the joined faces of the silicon nitride pipes. When the voids were not observed in the joint region, the border with the parent region was not clearly distinguishable (Fig. 1 (c)). This indicates that the parent silicon nitride directly bonded with each other. In our previous work, the gap between the silicon nitride pipes with rough or uneven joined faces was well filled by using melted insert material and the joined silicon nitride pipe was successfully fabricated without voids in the joint region[3,4]. Hence, when silicon nitride with the rough or uneven joined faces is used for parent material, the use of insert material will be required to obtain silicon nitride joints without forming voids in the joint region.

Table 1 shows flexural strength of the samples joined without using insert material at 1600°C. The joined samples indicated low strength value of about 170 MPa, which is much lower than the average strengths of the parent silicon nitride bulks (737 MPa) and the silicon nitride samples joined using Si_3N_4-Y_2O_3-Al_2O_3-SiO_2 powder mixture as insert material (about 680 MPa)[3,4]. On six tested samples joined without using insert material, all samples fractured from the joint region. In our previous work, out of six tested samples joined using the insert material at 1600°C, four was failed from the parent region[4]. The low joining strength and the fracture from the joint region in this work are due to the existence of voids in the whole joint region as shown in Fig. 1.

CONCLUSION

The two silicon nitride pipes of 1 m in length were joined without using insert material to fabricate the long silicon nitride pipe by a local-heating joining technique. The pipes had rough or uneven joint surface. The microstructure of the joint region was observed by SEM and the joining strength of the joined samples was measured using a four-point bending test. The SEM observations showed that voids were formed in the whole joint region. When no voids were observed in the joint region, the joint region was not clearly distinguished from the parent one. Flexural strength of the samples cut from the joined silicon nitride pipe indicated the average value of about 170 MPa. On six tested samples, all samples fractured from the joint region.

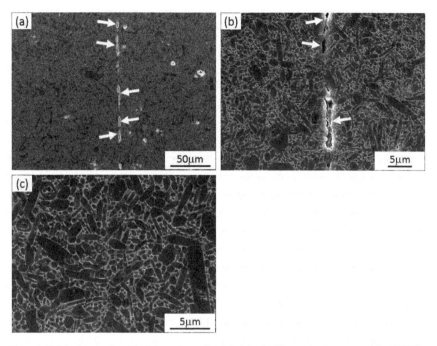

Figure 1. SEM micrographs of the joint regions of sample joined without using insert material at 1600°C.

Table 1. Room-temperature flexural strength of sample joined without insert material at 1600°C and parent silicon nitride bulk.

Sample	Flexural strength (MPa)
Joined silicon nitride	169 ± 52 (Max.: 264, Min.: 123)
Parent silicon nitride bulk	737 ± 46

ACKNOWLEDGMENT
This work was supported by the Project for "Innovative Development of Ceramics Manufacturing Technologies for Energy Saving" from the Ministry of Economy, Trade and Industry (METI) and New Energy and Industrial Technology Development Organization (NEDO).

REFERENCES
1) H. Kita, H. Hyuga, N. Kondo and T. Ohji, Exergy Consumption through the Life Cycle of Ceramic Parts, *Int. J. Appl. Ceram. Tech.*, **5**, 373-381 (2008).
2) H. Kita, H. Hyuga and N. Kondo, Stereo Fabric Modeling Technology in Ceramics Manufacture, *J. Eur. Ceram. Soc.*, **28**, 1079-1083 (2008).
3) M. Hotta, N. Kondo and H. Kita, Joining of Silicon Nitride Long Pipe by Local Heating, *Ceram. Eng. Sci. Proc.*, **32**, 89-92 (2011).
4) M. Hotta, N. Kondo, H. Kita, T. Ohji and Yasuhisa Izutsu, Joining of Silicon Nitride by Local Heating for Fabrication of Long Ceramic Pipes, *Int. J. Appl. Ceram. Tech.*, (in press), DOI: 10.1111/j.1744-7402.2012.02853.x

INTERFACIAL CHARACTERIZATION OF DIFFUSION-BONDED MONOLITHIC AND
FIBER-BONDED SILICON CARBIDE CERAMICS

H. Tsuda[1], S. Mori[1], M. C. Halbig[2] and M. Singh[3]
[1]Graduate School of Engineering, Osaka Prefecture University, Osaka, Japan
[2]NASA Glenn Research Center, Cleveland, Ohio, USA
[3]Ohio Aerospace Institute, NASA Glenn Research Center, Cleveland, Ohio, USA

ABSTRACT
    Diffusion bonding was used to join silicon carbide (SiC) to SiC substrates using 10 μm thick titanium interlayers. Two types of substrate materials were used: chemical vapor deposited (CVD)-SiC and SA-Tyrannohex™ (SA-THX); the latter has microstructures consisting of SiC fibers and a carbon layer. Microstructures of the phases formed during diffusion bonding were investigated with transmission electron microscopy (TEM) and selected-area diffraction (SAD) analysis. Attention was paid throughout to the direction of the SiC fibers in SA-THX with respect to the Ti interlayer. For example, the width of bonding region was 9 or 15 μm depending on whether the SiC fibers were parallel or perpendicular to the Ti interlayer, respectively. From the SAD pattern analysis, Ti_3SiC_2, $Ti_5Si_3C_x$, and $TiSi_2$ were identified in all samples. TiC and unknown phases appeared when a monolithic CVD-SiC substrate was used with a SA-THX substrate with SiC fibers perpendicular to the Ti interlayer. A high concentration of Ti_3SiC_2 and a lower concentration of $Ti_5Si_3C_x$ were formed when SA-THX with SiC fibers parallel to the Ti interlayer was processed. In contrast, a lower concentration of Ti_3SiC_2 phases and a higher concentration of $Ti_5Si_3C_x$ were formed with fibers perpendicular to the Ti interlayer. These differences are caused by the presence of the hexagonal carbon layer on the SiC fiber surface that acts like a grain boundary at the Ti interlayer. It appears that this layer retards the migration of Si and C atoms into the Ti interlayer during diffusion bonding.

INTRODUCTON
    Silicon carbide (SiC) is a very promising material for extreme environmental applications because of its excellent high temperature mechanical properties, oxidation resistance, and thermal stability. Advanced SiC-based ceramics and ceramic matrix composite materials are being developed for numerous aerospace and energy applications.[1] However, robust ceramic joining and integration technologies are needed before the ceramics can be implemented. The most commonly used joining methods include reaction bonding[2-4] and brazing.[5-6] Diffusion bonding techniques have also been utilized and hold significant promise.
    However, to have diffusion bonds with good thermomechanical properties, detailed knowledge of the phases formed during the reaction process is necessary. For example, Gottselig et al.[7] and Naka et al.[8] have reported on the bonding of SiC with Ti and the subsequent phases and diffusion path in the ternary system. However, the processing conditions varied and the microstructural analysis was not presented in detail. Previously, we reported in detail about various phases that form in the bonded area during diffusion bonding of SiC to SiC using scanning electron microscopy (SEM), X-ray diffraction (XRD) analysis, and energy dispersive spectroscopy (EDS).[9-11] Recently, we used transmission electron microscopy (TEM) for the microstructural evaluation of the phases formed during diffusion bonding with Ti as the interlayer.[12, 13]
    Ishikawa et al. have developed a tough ceramic (SA-Tyrannohex™ referred to as SA-THX) that consists of a highly ordered, closed-packed structure of fine hexagonal columnar fibers composed of crystalline β-SiC with a thin interfacial carbon layer between the fibers.[14, 15] In the present work, SA-THX was diffusion bonded to SA-THX with a Ti foil interlayer to examine the bonding effect of the SiC fiber orientation relative to the joining plane. We also present a detailed TEM microstructural

analysis of the phases formed during diffusion bonding, as well as selected area diffraction (SAD) analysis of samples prepared with a focused ion beam (FIB). The results are compared with those obtained from diffusion bonding of chemical vapor deposited-SiC (CVD-SiC) to CVD-SiC.

EXPERIMENTAL

CVD β-SiC substrates and SA-THX were obtained from Rohm & Hass (Woburn, MA, USA) and Ube Industries (Ube, Japan), respectively. The SA-THX is a fiber-reinforced composite with SA-Tyranno fiber bundles woven into an eight-harness satin weave that results in fibers being oriented in the transverse and longitudinal directions. Ti foil for 10 μm joining was obtained from Goodfellow Corporation (Glen Burnie, MD, USA). Before joining, all materials were ultrasonically cleaned in acetone for 10 min. Joints for both types of SiC were diffusion bonded at 1200°C with a pressure of 30 MPa. Joint processing was conducted in vacuum for 2 h (CVD-SiC) or for 4 h (SA-THX) at the peak temperature under load, followed by slow cooling at a rate of 2°C per min.

SEM observation and elemental analysis were performed with a Hitachi S-4700-I. All samples for TEM (200kV, JEOL, JEM-2000FX) were prepared by a FIB (FEI, Quant 3D), which allows us to obtain a precisely selected, clean, and less-damaged thin specimen from the diffusion bonded area, as described previously.[16]

RESULTS AND DISCUSSION

SEM Microstructure of Diffusion Bonded Samples

Fig. 1 shows a back-scattered electron image of diffusion bonded CVD-SiC using a 10-μm thick Ti foil, and Table I lists the compositions of the various phases in the joint that are labeled A, B, and C in Fig. 1. The composition of phase A is 51C-14Si-35Ti and that of phase B is 38C-27Si-35Ti. Phase C (37C-43Si-20Ti) has a relatively high Si concentration, indicating that it may be more ductile. Only a few insignificant microcracks were observed in this sample.

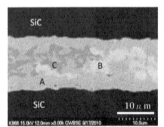

Table I. Composition of phases marked in Figure 1.

Spot	C at %	Si at %	Ti at %
A	51	14	35
B	38	27	35
C	37	43	20

Figure 1. Back-scattered electron image of diffusion-bonded CVD-SiC.

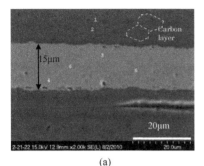 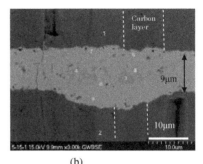

(a)   (b)

Figure 2. SEM images of diffusion bonds (a) Fiber direction is parallel to Ti foil,
(b) Fiber direction is perpendicular to Ti foil.

In Fig. 2(a) and (b), secondary and back-scattered electron images, respectively, of bonds are shown for SA-THX fibers set parallel (a) or perpendicular (b) to the 10 μm thick Ti foil that was the diffusion-bonded interlayer. Adequate diffusion bonds were obtained in all samples. In Fig. 2(a) and (b), hexagonal and straight carbon layers can be seen on the SA-THX surface. The morphology of the interface between the bond and the SA-THX with fibers set parallel to the Ti interlayer is smoother than those set perpendicular to the Ti. The 15 μm width of the diffusion bonds for SA-THX fibers set parallel to the Ti interlayer is wider than that (9 μm) of perpendicular fiber SA-THX substrate. In addition, in Fig. 2(b), where the fiber is perpendicular to the Ti layer, a small microcrack can be seen.

Table II. Compositions of phases.
(Fiber direction is parallel to Ti foil)

Spot	C at %	Si at %	Ti at %	Probable phase
1	54.28	45.72		SiC
2	59.52	40.48		
3	44.89	15.79	39.32	Ti$_3$SiC$_2$
4	44.43	15.65	39.92	
5	-	69.39	30.61	TiSi$_2$
6	-	66.66	33.34	

Table III. Compositions of phases.
(Fiber direction is perpendicular to Ti foil)

Spot	C at %	Si at %	Ti at %	Probable phase or common and unique phase
1	57.22	42.78	-	SiC
2	54.96	45.04	-	
3	3.21	15.23	81.56	unknown
4	15.27	21.83	62.90	Ti$_3$SiC$_2$
5	11.02	22.44	66.54	
6	11.49	4.82	83.69	unknown
7	12.27	4.75	82.98	
8	12.83	4.63	82.54	
9	12.18	4.73	83.09	

In Tables II and III, the compositions are listed for all phases identified in the diffusion bonds (Fig. 2), as determined by EDS. The phases in Fig. 2(a) were identified as SiC, Ti$_3$SiC$_2$, and TiSi$_2$. In Fig. 2(b), phases SiC and Ti$_3$SiC$_2$ were identified.

TEM of Diffusion-Bonded Samples

TEM imaging revealed details in the microstructures of the phases found in the three different diffusion-bonded samples. Average grain sizes of SiC prepared by CVD (data not shown) range from

3–5 μm. Furthermore, twin spots with streaks in the SAD patterns indicated stacking faults or micro-twins located in crystalline SiC. In contrast, average grain sizes of SiC in SA-THX are from 0.2–0.5 μm, and SAD clearly indicates that they are crystalline β-SiC. Twin spots with streaks can also be seen in SA-THX. The diffusion bonds consisted of many small grains with lengths of 2–4 μm, and widths of 1–2 μm.

Fig. 3(a) and (b) shows TEM micrographs of SA-THX fibers set parallel and perpendicular to the Ti foil, respectively. SAD patterns were taken at the labeled locations and the probable phases are summarized in Table IV. The detailed phase identification and SAD pattern analyses were conducted on more than 20 grains for all three samples (individual phases are identified in the micrographs and are listed in corresponding tables). Table V gives the calculated percentage content of each phase. Figs. 4 and 5 show SAD patterns of individual phases of SA-THX fibers set parallel and perpendicular to the Ti foil, respectively. TEM analysis indicates that diffusion bonds in samples with parallel SA-THX fibers consist of Ti_3SiC_2, $Ti_5Si_3C_x$, and $TiSi_2$ phases. The fractions of the phases formed during diffusion bonding are 84.2% Ti_3SiC_2, 5.3% $Ti_5Si_3C_x$, and 10.5% $TiSi_2$. In the Ti-Si-C ternary system, the identified phases are consistent with the results of Gottselig et al.[7] In this diffusion bond, the $Ti_5Si_3C_x$ phase was scarcely observed. Naka et al.[8] suggested that $Ti_5Si_3C_x$ is an intermediate phase that is not present when phase reactions have gone to completion. With XRD, Naka et al.[8] confirmed the presence of Ti_3SiC_2 and $TiSi_2$ in the final stage of processing, when Ti was used to join SiC to SiC. As mentioned above, Ti_3SiC_2 is the primary phase and the fractions of other phases are extremely small. This suggests that the reaction between the SA-THX with parallel fibers and the 10 μm Ti foil interlayer was complete.

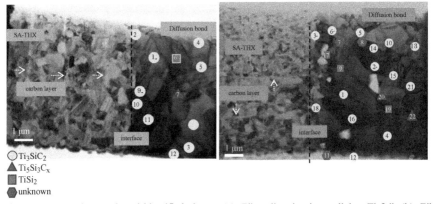

$\bigcirc$ Ti_3SiC_2
$\blacktriangle$ $Ti_5Si_3C_x$
$\blacksquare$ $TiSi_2$
$\blacklozenge$ unknown

Figure 3. TEM micrograph and identified phases. (a): Fiber direction is parallel to Ti foil, (b): Fiber direction is perpendicular to Ti foil

Table IV. Summary of probable phases determined from SAD patterns.

	CVD-SiC	SA-THX parallel to Ti foil	SA-THX perpendicular to Ti foil
1	Ti_3SiC_2	Ti_3SiC_2	Ti_3SiC_2
2	Ti_3SiC_2	Ti_3SiC_2	Ti_3SiC_2
3	Ti_3SiC_2	Ti_3SiC_2	Ti_3SiC_2
4	Ti_3SiC_2	Ti_3SiC_2	Ti_3SiC_2
5	$Ti_5Si_3C_x$	Ti_3SiC_2	Ti_3SiC_2
6	$Ti_5Si_3C_x$	$TiSi_2$	Ti_3SiC_2
7	$Ti_5Si_3C_x$	$Ti_5Si_3C_x$	unknown
8	TiC	Ti_3SiC_2	unknown
9	Ti_3SiC_2	Ti_3SiC_2	$TiSi_2$
10	Ti_3SiC_2	Ti_3SiC_2	Ti_3SiC_2
11	$TiSi_2$	Ti_3SiC_2	unknown
12	Ti_3SiC_2	Ti_3SiC_2	$Ti3SiC_2$
13	Ti_3SiC_2	Ti_3SiC_2	Ti_3SiC_2
14	TiC	Ti_3SiC_2	Ti_3SiC_2
15	$TiSi_2$	Ti_3SiC_2	Ti_3SiC_2
16	Ti_3SiC_2	$TiSi_2$	Ti_3SiC_2
17	Ti_3SiC_2	Ti_3SiC_2	$TiSi_2$
18	Ti_3SiC_2	Ti_3SiC_2	Ti_3SiC_2
19	Ti_3SiC_2	Ti_3SiC_2	$TiSi_2$
20	Ti_3SiC_2	Ti_3SiC_2	Ti_5Si3C_x

	CVD-SiC	SA-THX parallel to Ti foil	SA-THX perpendicular to Ti foil
21	unknown	Ti_3SiC_2	Ti_3SiC_2
22	unknown	Ti_3SiC_2	$Ti_5Si_3C_x$
23	$Ti_5Si_3C_x$	-	-
24	Ti_3SiC_2	-	-
25	Ti_3SiC_2	-	-
26	Ti_3SiC_2	-	-
27	Ti_3SiC_2	-	-

Table V. Calculation of phases formed in diffusion bonding process.

	CVD-SiC	SA-THX parallel to Ti foil	SA-THX perpendicular to Ti foil
Ti_3SiC_2	63.5	84.2	63.6
$Ti_5Si_3C_x$	18.2	5.3	9.1
$TiSi_2$	6.1	10.5	13.6
TiC	6.1	0	0
unknown	6.1	0	13.6
Total (%)	100	100	100

Fig. 3(b) is a TEM micrograph taken from SA-THX fibers set perpendicular to the Ti foil; identified phases (Ti_3SiC_2, $Ti_5Si_3C_x$, and $TiSi_2$) are shown. An unidentified phase was also detected. Fractions of the known phases are 63.6% Ti_3SiC_2, 9.1% $Ti_5Si_3C_x$, and 13.6% $TiSi_2$. Compared with the SA-THX fibers set parallel to the Ti foil, the fraction of Ti_3SiC_2 is smaller and the fraction of $Ti_5Si_3C_x$ is considerably greater. The reactions between SiC and Ti were complete for a 4 h hold when the SA-THX fiber was set parallel to Ti foil. In contrast, the reaction for the SA-THX fiber set perpendicular to foil was incomplete when processed at the same temperature and hold time.

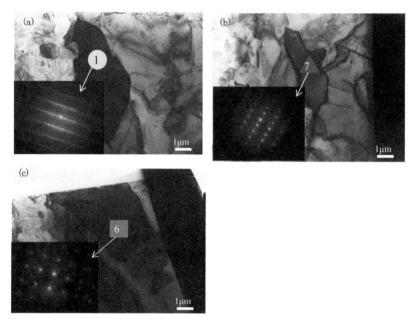

Figure 4. Representative TEM micrographs and SAD patterns of a diffusion bond.
(Fiber direction is parallel to Ti foil)
(a) Ti_3SiC_2 (B=[11-20]), (b)$Ti_5Si_3C_x$ (B=[411]=[-72-53]) and (c)$TiSi_2$ (B=[102])

Effect of SiC Fiber Orientation on the Reaction Products

In the SA-THX specimens with the fibers set parallel to joints, stable phases of Ti_3SiC_2 and $TiSi_2$ were predominantly formed. Thus, the chemical reaction between the SiC of the SA-THX substrate and the Ti was complete, and this geometrical combination had no microcracks. In the joint formed with the perpendicular SA-THX fibers, a relatively large number of detrimental, average-sized $Ti_5Si_3C_x$ grains were detected. According to the literature,[7] $Ti_5Si_3C_x$ is a solid solution of Ti_5Si_3 containing a small atomic percentage of carbon. In addition, this intermediate phase undergoes anisotropic thermal expansion.[17, 18] As the joint cools after processing, mismatches in the coefficients of thermal expansion induce thermal stress that could explain microcrack formation in the perpendicular SA-THX fiber specimens.

As discussed above, there are high concentrations of Ti_3SiC_2 phases and low concentrations of $Ti_5Si_3C_x$ phases when SA-THX SiC fibers are parallel to the Ti interlayer. In contrast, less of the stable phase Ti_3SiC_2 is formed and more of the $Ti_5Si_3C_x$ phase is formed in SA-THX fibers set perpendicular to the interlayer. A schematic view of fiber direction in SA-THX and the geometrical relation between the carbon layer and the Ti foil is shown in Fig. 6. The basal plane of the carbon layer in the hexagonal columns always faces the Ti interlayer, just like a grain boundary, when the SiC fibers of SA-THX are perpendicular. This probably retards the mobility of Si and C atoms that are migrating into the Ti interlayer during diffusion bonding. However, Si and C atoms can migrate readily when the SA-THX fibers are set parallel to the Ti layer, because the prismatic plane of the carbon layer in the hexagonal columns commonly does not face the Ti foil. This enhances the mobility of Si and C atoms into the Ti interlayer during diffusion bonding.

Figure 5. Representative TEM micrographs and SAD patterns of a diffusion bond.
(Fiber direction is perpendicular to Ti foil)
(a)Ti$_3$SiC$_2$ (B=[11-20]), (b)Ti$_5$Si$_3$Cx (B=[121]=[01-11]), (c)TiSi$_2$ (B=[102]) and (d) unknown phase

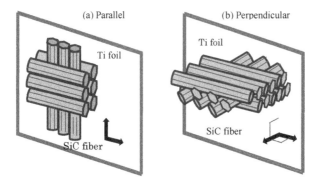

Figure 6. Schematic view of fibers in SA-THX relative to the Ti foil.

CONCLUSION

Two different SiC substrates (CVD-SiC and SA-THX) were diffusion bonded using a 10 μm thick Ti interlayer. After diffusion bonding at 1200°C, the microstructure of the bonded region was revealed by TEM. The results are summarized as follows.

(1) It was found that well-diffused bonds were formed in all samples. In the SA-THX fibers set perpendicular to the Ti interlayer, an insignificant amount of microcracking was observed.

(2) Ti_3SiC_2, $Ti_5Si_3C_x$, and $TiSi_2$ phases were identified in all samples. Furthermore, TiC and unknown phases appeared in the samples where the CVD-SiC and the SA-THX fibers set perpendicular to the Ti foil were diffusion bonded. Additionally, Ti_3SiC_2 was formed more, and $Ti_5Si_3C_x$ was formed less, when SA-THX fibers were set parallel to the Ti interlayer.

(3) The width of the bonded region was 9 μm when the fibers were perpendicular and was 15 μm when the fibers were parallel. Based on the analysis of SAD patterns, Ti_3SiC_2, $Ti_5Si_3C_x$, and $TiSi_2$ were identified in all samples. TiC and unidentified phases also appeared when a monolithic CVD-SiC substrate and SA-THX SiC with fibers set perpendicular to the Ti interlayer were used.

(4) Diffusion bonds had Ti_3SiC_2 phases in high concentrations and $Ti_5Si_3C_x$ in low concentrations when SA-THX had fibers set parallel to the Ti interlayer. However, stable phase Ti_3SiC_2 formed less and $Ti_5Si_3C_x$ formed more in diffusion bonds for SA-THX fibers set perpendicular to the Ti interlayer. This is most likely due to the presence of the hexagonal carbon layer that always faces the Ti interlayer, and which retards the mobility of Si and C atoms migrating into the Ti interlayer during diffusion binding.

ACKNOWLEDGEMENTS

Hiroshi Tsuda would like to thank Mrs. Taeko Yuki and Mr. Tsukasa Koyama of Osaka Prefecture University for preparing TEM samples using a FIB.

REFERENCES
[1]P. J. Lamicq, G. A. Bernhart, M. M. Dauchier, and J. G. Mace, SiC/SiC Composite Ceramics, *Am.*

Ceram. Soc. Bull., **65**[2], 336–338 (1986).

[2]M. Singh, A Reaction Forming Method for Joining of Silicon Carbide-based Ceramics, *Scr. Mater.*, **37**, Issue 8 1151-1154 (1997).

[3]M. Singh, Joining of Sintered Silicon Carbide Ceramics for High Temperature Applications, *J. Mater. Sci. Lett.*, **17**, Issue 6, 459-461 (1998).

[4]M. Singh, Microstructure and Mechanical Properties of Reaction Formed Joints in Reaction Bonded Silicon Carbide Ceramics, *J. Mater. Sci.*, **33**, 1-7 (1998).

[5]V. Trehan, J.E. Indacochea, and M. Singh, Silicon carbide brazing and joint characterization, *J. Mech. Behav. Mater.*, **10**, Issue 5-6 341-352 (1999).

[6]M.G. Nicholas, Joining Processes: Introduction to Brazing and Diffusion Bonding, *Kluwer Academic Publishers*, Dodrecht, 1998.

[7]B. Gottselig, E. Gyarmati, A. Naoumidis, and H. Nickel, Joining of Ceramics Demonstrated by the Example of SiC/Ti, *J. Eur. Ceram. Soc.*, **6**, 153-160 (1990).

[8]M. Naka, J. C. Feng, and J. C. Schuster, Phase Reaction and Diffusion Path of the SiC/Ti System, *Metall. Mater. Trans. A*, **28A**, 1385-1390 (1997).

[9]M. Singh and M.C. Halbig, Bonding and Integration of Silicon Carbide Based Materials for Multifunctional Applications, *Key Eng. Mater.*, **352**, 201-206 (2007).

[10]M.C. Halbig and M. Singh, Development and Characterization of the Bonding and Integration Technologies Needed for Fabricating Silicon Carbide-based Injector Components in *"Advanced Processing and Manufacturing Technologies for Structural and Multifunctional Materials II"* edited by T. Ohji and M. Singh, CESP, Vol. **29**, Issue 9, pp. 1-14, Wiley, NY and American Ceramic Society (2009).

[11]M.C. Halbig and M. Singh, Diffusion Bonding of Silicon Carbide for the Fabrication of Complex Shaped Ceramic Components, " *Ceramic Integration and Joining Technologies: From Macro- to Nanoscale*, Eds. M. Singh, T. Ohji, R. Asthana and S. Mathur, John Wiley & Sons, 2011.

[12]M. C. Halbig, M. Singh and H. Tsuda, Integration Technologies for Silicon Carbide-Based Ceramics for Micro-Electro-Mechanical Systems-Lean Direct Injector Fuel Injector Applications, *"International Journal of Applied Ceramic Technology*, **9**, 677-687 (2012).

[13]H. Tsuda, S. Mori, M. C. Halbig and M. Singh, TEM Observation of the Ti Interlayer between SiC Substrates during Diffusion Bonding, Ceramic Engineering and Science Proceedings, **33**, Issue 8, 81-89 (2012).

[14]T. Ishikawa, S. Kajii, K. Matsunaga, T. Hogami, Y. Kohtoku and T. Nagasawa, A Tough, Thermally Conductive Silicon Carbide Composite with High Strength up to 1600°C in Air, *Science*, **282**, 1265-1297 (1998).

[15]T. Ishikawa, Y. Kohtoku, K. Kumagawa, T. Yamamura and T. Nagasawa, "High-strength Alkali-Resistant Sintered SiC Fibre Stable to 2200°C, *Nature*, **391**, 773-775 (1998).

[16]J. Ayanne, L. Beaunier, J. Boumendil, G. Ehret and D. Laub, *Sample Preparation Handbook for Transmission Electron Microscopy: Techniques*, 1st Edition, P. 135, Springer, N.Y., 2010.

[17]J. H. Schneibel and C. J. Rawn, Thermal Expansion Anisotropy of Ternary Silicides Based on Ti_5Si_3, *Acta Mater.*, **52**, 3843-3848 (2004).

[18]L. Zhang and J. Wu, Thermal Expansion and Elastic Moduli of the Silicide Based Intermetallic Alloys $Ti_5Si_3(X)$ and Nb_5Si_3, *Scr. Mater.*, **38**, Issue 2 307-313 (1998).

ROUND ROBIN ON INDENTATION FRACTURE RESISTANCE OF SILICON CARBIDE
FOR SMALL CERAMIC PRODUCTS

Hiroyuki Miyazaki, Yu-ichi Yoshizawa
National Institute of Advanced Industrial Science and Technology (AIST)
Anagahora 2266-98, Shimo-shidami, Moriyama-ku, Nagoya 463-8560, Japan

Kouichi Yasuda
Tokyo Institute of Technology
2-12-1-S7-14, Ookayama, Meguro-ku, Tokyo 152-8552, Japan

ABSTRACT

Round robin on indentation fracture resistance, K_{IFR} of silicon carbides was conducted by thirteen laboratories in order to assess the reliability of the IF method. The reported K_{IFR} varied widely from 3.43 to 4.43 MPa·m$^{1/2}$ at the first round robin, where the crack length was measured mainly with a normal optical microscope equipped with the <u>hardness</u> tester. Our re-measurements of the same indentations which were returned to the authors clarified that the crack lengths themselves were constant and that the measuring errors of the crack length reported by each participants were in the range of 55 to 115 μm, which was the main origin of the wide scatter of K_{IFR}. Powerful microscopy with both an objective lens of 40x and a traveling stage was employed at the second round robin for the correct measurements of crack length. K_{IFR} obtained by this technique exhibited a consistent value of 3.23 ± 0.10 MPa·m$^{1/2}$ and was much closer to that re-measured by authors. It was revealed that the high resolving power of the objective lens of 40x enabled to find exact crack tips easier, which resulted in the good matching of K_{IFR} between laboratories. It was suggested that the observation of indentations with powerful optics was effective for improving the reproducibility of the IF method.

INTRODUCTION

Many small ceramic products and components such as bearing balls and cutting tools are used world widely.[1] Evaluation of the fracture toughness from such a small parts themselves is necessary for the quality control and/or the classification of the grade of products. However, standard toughness tests such as single edge-precracked beam (SEPB)[2, 3] and surface flaw in flexure (SCF) methods[4] are difficult to apply since they require test specimens larger than these products. One of the alternative tests to measure the fracture toughness of small ceramic parts is the indentation fracture (IF) method, which has been widely used since it has been proposed by Lawn and his co-workers.[5] This method is particularly useful when the sizes of available specimens are limited. However, there has been rigorous arguments that the value measured by the IF method does not represent the real fracture toughness and that the term "indentation fracture resistance, K_{IFR}" should be used when the IF method was applied.[6, 7] Thus, the American standard specification for silicon nitride bearing balls adopts the term "indentation fracture resistance, K_{IFR}" for the apparent fracture toughness measured by the IF method.[8]

The other serious problem of the IF method is the poor between-laboratory consistency, which was revealed by round-robin tests conducted in order to standardize the indentation fracture test for ceramics (e.g. VAMAS, [9-11] etc.[12]) about two decades ago. However, it is likely that the reproducibility of the IF test is improved as compared with those reported by previous round-robin tests since processing of structural ceramics has made a big progress during the decades and a measuring instrument has been refined as well. In our previous study, accuracy of the IF test was checked for silicon nitrides by an international round-robin test with six laboratories.[13] An excellent consistency of K_{IFR} between laboratories was attained when bearing-grade silicon nitrides were used.[13]   It was also revealed that the error in reading the crack length was diminished by using a powerful microscope.

In this study, the reliability of the IF test for silicon carbides was investigated by the round robin with thirteen laboratories in Japan.   The participants consisted of six universities, five companies and two national laboratories.   The round-robin test was conducted twice with the same indented samples but by the different measuring methods.   At the first round robin, almost all labs used an optical microscope with a low magnification of 100x – 280x.   After the measurements by each lab, the test specimens were returned to the authors and their indentations were re-measured with a powerful optics in order to find out which was the origin of the scatter of K_{IFR}, that is, the difference of the real size of indentations itself or the systematic biases of measurement due to different operators.   It is well known that slow crack growth does not occur in SiC densified with B additions since they typically do not leave residual phase along grain boundaries.   Then, the re-measured crack lengths were compared directly with those reported values.

In the second round-robin test, almost all of the participants observed the indentations at high magnifications of 400x or higher by using an objective lens of 40x.   The distances of crack tips were determined from the shift of the stage of the microscope since the cracks extended over the range of the microscope.   The effect of the powerful objective lens on the reproducibility of the results was studied and discussed with conjunction of its resolving power.

EXPERIMENTAL PROCEDURE

Materials

Commercially available silicon carbide ceramics (IBICERAM SC-850, IBIDEN Co., Ltd.) sintered with B and C as sintering additives were used in this study.   The bulk density of the sample was 3.024 g/cm^3 and the relative density was calculated to be 94.0% by using the theoretical density of 3.217 g/cm^3.   The Young's modulus obtained by the ultrasonic pulse echo method was 365 GPa.   Rectangular specimens with dimensions of 4 mm x 3 mm x 38 mm were machined from the sintered samples.   The larger 4 mm x 38 mm surface was polished to a mirror finish for indentations.

Test procedure

Each laboratory made more than eight Vickers indentations with a hardness tester.   The indentation load was 196 N for labs nos. 1 – 10 and 98 N for labs nos. 11 – 13.   The indentation contact time was 15 s.   The lengths of the impression diagonals, 2a, and surface cracks, 2c, were measured immediately after the indentation.   Only indentations whose four primary cracks emanated straight forward from each corner were accepted.   Indentations with badly split cracks or with gross chipping were rejected as well as those whose horizontal crack length differed by more than 10% from the vertical one.

Eleven labs out of thirteen used a total magnification of 100x – 280x and two labs (lab nos. 3 and 10) employed much higher magnification of 400x and 500x at the first round robin. In order to investigate the effect of the low magnification on the measurements, the data from the former labs were analyzed in this paper.   The microscopes of nine labs were furnished with their testers and the magnifications of the objective lens were 10x or 13x.   Six labs out of nine observed impressions directly with an eyepiece of 10x, so that the total magnification was 100x or 130x.   Lab nos. 9, 12 and 13 attached both a CCD camera and a monitor to the microscope instead of a eyepiece, which resulted in a little bit higher magnification of 250x or 280x.   The measurements with a measuring microscope or a metallurgical microscope were performed by lab nos. 4 and 11 and the total magnifications were 100x.

The indentation fracture resistance, K_{IFR}, was determined from the as-indented crack lengths by the Niihara's equation for the median crack system as follows: [14]

$$K_{IFR} = 0.0309(E/H)^{2/5}Pc^{-3/2} \qquad (1)$$

where E and H are Young's modulus and the Vickers hardness, respectively, P is the indentation force, and c is the half-length of as-indented surface crack length. In this study, Young's modulus mentioned above was used. K_{IFR} was calculated for each indentation using the hardness value obtained for each impression. Those calculated K_{IFR} together with the raw data were collected to the test organizer.

All these samples indented by each lab were returned to author's laboratory (AIST) to re-measure the sizes of indentations. It was deemed that a good resolution could be obtained by a measuring microscope, by which the crack tips are detected at the high magnification of 500x and the spacing between the tips is measured preciously by traveling the stage with a readout resolution of 1 μm. Thus, a measuring microscope with an objective lens of 50x (total magnification: 500x) was employed for the re-measurements. By comparing those data measured by each lab and AIST, it was revealed which factor was dominant, that is, whether the real crack lengths themselves differed due to the variation in hardness tester or the bias of reading the length affected the calculated K_{IFR}.

K_{IFR} reported by each lab exhibited poor consistency and was different from those re-measured value by the authors, implying the detection of exact crack tip was difficult with the poor microscope. It is natural to suppose that higher magnification of the object lens would be helpful to find crack tips. Then, the second round-robin test using a microscope equipped with both a powerful objective lens and a traveling stage was conducted in order to improve the accuracy of the measurements. The indented samples were sent to twelve laboratories again and the indentations made in the first round-robin test were re-measured using both an objective lens of 40x and a traveling stage. Half of labs observed the impressions directly at a total magnification of 400x with the eyepiece of 10x . A CCD camera and a monitor were employed by the rest of labs to obtain total magnifications of 600x or higher.

RESULTS AND DISCUSSION

First round robin using a low magnification

The crack lengths, $2c$, measured at a low total magnification of 100x – 280x were plotted in figure 1 (closed square). The variation of the crack lengths of indentations at 196 N was very wide (574 – 657 μm). The scattering of $2c$ of indentations at 98 N was also large. The crack lengths re-measured by authors with the measuring microscope are also shown as open triangles in figure 1. Almost constant and much longer $2c$ values of ~ 700μm were observed among the laboratories for the indentations at 196 N (lab. nos. 1 – 10), indicating that real $2c$ themselves were hardly affected by the different indenters as reported by authors previously.[15] The comparison of $2c$ between the each lab's value and our re-measured one revealed that all the participants missed the real crack tips when they employed the low magnification and that their misreading of $2c$ ranged from 55 to 115 μm. The same tendency was also obtained for the indentations at 98 N. It was inferred that detecting exact crack tips at the low magnification were difficult and susceptible to the subjectivity of the operators.

By contrast, the diagonal sizes, $2a$, of each lab's indentations did not vary so widely as those of crack lengths and their deviations from those re-measured by authors were not so significant. For example, $2a$ indented at 196 N were in the range of 133 – 151μm, whereas re-measured $2a$ resided around 142 μm.

The ratios of c/a for all the impressions were more than 2.5 regardless of the indentation load, suggesting that the crack system was the median type. Then, indentation fracture resistance, K_{IFR} were calculated with equation 1 and are shown in figure 2. Large scatter of K_{IFR} $(3.43 - 4.43 \, \text{MPa} \cdot \text{m}^{1/2})$ was observed among the laboratories (closed square), while re-measured K_{IFR} by authors exhibited the constant values of $3.11 \pm 0.10 \, \text{MPa} \cdot \text{m}^{1/2}$ (open triangle). It is obvious that the each lab's K_{IFR} was not reliable and that the major origin of such uncertainty was the error in reading crack length.

Second round robin using both an objective lens of 40x and a traveling stage
In order to improve the accuracy of the measurements of crack length, the objective lens with a high magnification of 40x and the traveling stage were employed at the second round robin. SiC samples were sent back again to each laboratory and the same indentations made at the first round robin were measured by each lab. The raw data of crack lengths is presented in figure 3 (closed circle). All crack lengths resided between 662 and 701 μm for the indentations at 196 N when measured by this technique and its standard deviation was only 18 μm, which was markedly improved as compared with that of the first round robin at the low magnification. The difference of $2c$ between the each lab's value and our re-measured one became much smaller than that of the first round robin, indicating that the misreading of $2c$ for the participants decreased significantly to ~ 20 μm. The same could be said for the indentations at 98 N as well. The diagonal sizes reported by each lab were the same as those of the first round robin and a relatively small variation in $2a$ was attained again. Figure 4 shows that K_{IFR} (closed circle) obtained by using both an objective lens of 40x and a traveling stage varied little among the laboratories and were $3.23 \pm 0.10 \, \text{MPa} \cdot \text{m}^{1/2}$, which matched well with the values re-measured by authors (open triangle). Thus the reproducibility of this method could be regarded as adequate. All these things make it clear that the scatter of the crack length could be diminished by using both the objective lens of 40x and the traveling stage and that the accuracy of K_{IFR} of SiC could be improved significantly by this technique.

Effect of the magnification of the objective lens on the precision of the crack-tip detection
The resolving power of an optical microscope is determined by the performance of an objective lens as follows:

$$\delta = 0.61 \, \lambda \, / \, NA \quad\quad\quad (2)$$

where δ is the two point resolving power, λ is the wave length of the light and NA is the numerical aperture of the objective lens. The NA of the objective lens of 50x and 10x used in our laboratory were 0.55 and 0.2, respectively. The resolving power for the objective lens of 50x and 10x was calculated to be 0.55 μm and 1.5 μm, respectively when green light (λ = 500 nm) was used. The improved precision by using the objective lens of 50x can be explained by its high resolving power as follows. Observers usually start searching the crack tip from the corner of the impression where the crack is wide. The identification of crack becomes progressively difficult as observers approach to the crack tip since the crack opening displacement gradually decreases. When the crack width became less than 1.5 μm, the operators can't identify the crack clearly if the objective lens of 10x is used because of its limited resolving power. Then the crack with width less than 1.5 μm would be missed and the crack length would be measured shorter. By contrast, the very narrow crack with width of 0.6 μm in the vicinity of the real crack tip can be detected with the objective lens of 50x. Accordingly, the high resolution image produced by objective lens of 50x enable to find the real crack tip easier and can give more accurate crack length than that obtained with the objective lens of 10x. It is deemed that higher magnification of objective lens of 40x or higher should be used for the reliable measurement of indentation fracture resistance.

CONCLUSIONS

In order to investigate the reproducibility of the indentation fracture resistance, K_{IFR}, domestic round robin tests were performed twice using SiC sintered with B and C.

(1) K_{IFR} varied widely when the normal microscope equipped with the tester was employed at the first round-robin test.

(2) Re-measurement of the returned sample by the authors with the measuring microscope at a total magnification of 500x indicated that the crack lengths themselves were constant and that the error in measuring the crack length was dominant origin of the large scatter of K_{IFR}.

(3) The scatter of K_{IFR} among the laboratories was reduced greatly when the object lens with the high magnification of 40x and the traveling stage were used at the second round robin test. The improvement was attributed to the exact identification of the crack tips by the high magnification.

(4) It was revealed that the reproducibility of K_{IFR} of SiC could be improved significantly by employing the powerful optics with the traveling stage for the crack length measurements.

ACKNOWLEDGEMENTS

The authors express sincere thanks to all the participants involved in this round-robin test. The authors also acknowledge the workshop for innovation in reliability of bulk ceramics in the ceramic society of Japan.

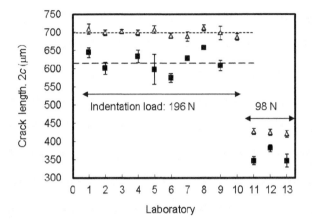

Figure 1. Crack length, 2c, of the SiC samples measured at the first round robin. Closed squares represent each laboratory's average observed at a lower magnification from 100x to 280x, while open triangles denote those measured by authors with a measuring microscope at a total magnification of 500x. One standard deviation (error bars) is also shown. The dashed line represents the average of all the reported data (196 N) at the lower magnification, whereas the dotted line is the averages of all the re-measured ones (196 N) by authors.

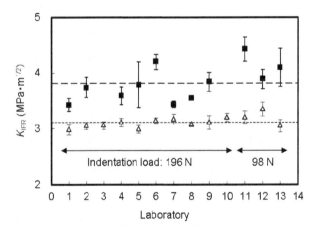

Figure 2. Results of the first round robin on indentation fracture resistance of the SiC samples. Closed squares represent each laboratory's average observed at a lower magnification from 100x to 280x, while open triangles denote those measured by authors with a measuring microscope at a total magnification of 500x. One standard deviation (error bars) is also shown. The dashed line represents the average of the reported data at the lower magnification, whereas the dotted line is the averages of the re-measured ones by authors.

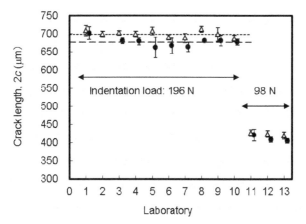

Figure 3. Crack length, 2c, of the SiC samples measured at the second round robin. Closed circles represent each laboratory's average observed at <u>total magnifications of 400x or higher</u>, while open triangles denote those measured by authors with a measuring microscope at a total magnification of 500x. One standard deviation (error bars) is also shown. The dashed line represents the average of all the reported data (196 N) by the participants, whereas the dotted line is the averages of all the re-measured ones (196 N) by authors.

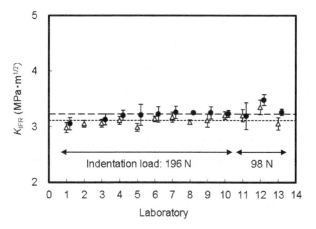

Figure 4. Results of the second round robin on indentation fracture resistance of the SiC samples. Closed circles represent each laboratory's average observed at a total magnification of 400x, while open triangles denote those measured by authors with a measuring microscope at a total magnification of 500x. One standard deviation (error bars) is also shown. The dashed line represents the average of the <u>reported data</u> at <u>total magnifications of 400x or higher</u>, whereas the dotted line is the averages of the re-measured ones by authors.

REFERENCES

[1] K. Komeya, Material development and wear applications of Si_3N_4 ceramics, *Ceram. Trans.*, **133**, 3-16 (2002).

[2] T. Nose, T. Fujii, Evaluation of fracture toughness for ceramic materials by a single-edge-precracked-beam method, *J. Am. Ceram. Soc.*, **71**, 328-333 (1988).

[3] Testing methods for fracture toughness of fine ceramics, Japanese Industrial Standard, JIS R 1607, 1995.

[4] Fine ceramics (Advanced ceramics, Advanced technical ceramics) – Determination of fracture toughness of monolithic ceramics at room temperature by the surface crack in flexure (SCF) method, International Organization for Standards, ISO 18756, Geneva, 2003.

[5] B.R. Lawn, A.G. Evans and B. Marshall, Elastic/plastic indentation damage in ceramics: the median/radial crack system, *J. Am. Ceram. Soc.*, **63**, 574-581 (1980).

[6] G.D. Quinn, Fracture toughness of ceramics by the Vickers indentation crack length method: a critical review, *Ceram. Eng. Sci. Proc.*, **27**, (2006).

[7] G.D. Quinn, R.C. Bradt, On the Vickers indentation fracture toughness test, *J. Am. Ceram. Soc.*, **90**, 673-680 (2007).

[8] Standard specification for silicon nitride bearing balls, ASTM F 2094/M2094M, 2008.

[9] D.M. Butterfield, D.J. Clinton and R. Morell, The VAMAS hardness round-robin on ceramic materials, VAMAS report#3, National physical laboratory, Teddington, Middlesex, United Kingdom, 1989.

[10] H. Awaji, T. Yamada and H. Okuda, Result of the fracture toughness test round robin on ceramics – VAMAS Project-, *J. Ceram. Soc. Jpn.*, **99**, 417-22 (1991).

[11] H. Awaji, J. Kon and H. Okuda, The VAMAS fracture toughness test round-robin on ceramics, VAMAS report#9, Japan fine ceramic center, Nagoya, Japan, 1990.

[12] Report of preliminary investigation for standardization of fine ceramics, Japanese fine ceramics association, Japan, 1988.

[13] H. Miyazaki, Y. Yoshizawa, K. Hirao and T. Ohji, Indentation fracture resistance test round robin on silicon nitride ceramics, *Ceram. Int.*, **36**, 899-907 (2010).

[14] K. Niihara, R. Morena and D.P.H. Hasselman, Evaluation of K_{1c} of brittle solids by the indentation method with low crack-to-indent ratios, *J. Mater. Sci. Lett.*, **1**, 13-16 (1982).

[15] H. Miyazaki, H. Hyuga, Y. Yoshizawa, K. Hirao, T. Ohji, Measurement of Indentation Fracture Toughness of Silicon Nitride ceramics, *Key Eng. Mater.*, **352**, 45-48 (2007).

NUMERICAL ANALYSIS OF MICROSTRUCTURAL FRACTURE BEHAVIOR IN NANO COMPOSITES UNDER HVEM

Hisashi Serizawa
Joining and Welding Research Institute, Osaka University
11-1 Mihogaoka, Ibaraki, Osaka 567-0047, Japan

Tamaki Shibayama
Faculty of Engineering, Hokkaido University
Kita 13, Nishi 8, Kita-ku, Sapporo, Hokkaido 060-8628, Japan

Hidekazu Murakawa
Joining and Welding Research Institute, Osaka University
11-1 Mihogaoka, Ibaraki, Osaka 567-0047, Japan

ABSTRACT
        A new nano-mechanics in-situ transmission electron microscope (TEM) experimental apparatus was developed for measuring applied load and indentation depth curve during the in-situ observation in high voltage electron microscope (HVEM).   By using this new apparatus, the crack initiation and propagation at the interface between SiC matrix and carbon layer coated on SiC fiber in SiC/SiC composite can be directly observed with measuring the load – displacement curve of a miniaturized double notch shear (DNS) test.   The inter-laminar shear strength of NITE (nano-powder infiltration and transient eutectic process) SiC/SiC composite was estimated as 2.8×10^3 MPa, which is about thirty times higher than the result obtained by the conventional DNS test.   From the finite element analysis with the interface element about the miniaturized DNS test, it was revealed that the maximum load can be predicted by assuming the higher yield stress of SiC and the reason of the extremely high shear strength seems to be this higher yield stress due to the lower density and distribution of defects.

INTRODUCTION
        Advanced multifunctional materials have been developed by controlling their microstructure precisely or by combining various dissimilar materials.   Prior to the practical usage of these materials, it is necessary to evaluate their complicate functions and to estimate the mechanical properties including the fracture strength.   In general, the macroscopic mechanical properties, such as Young's modulus, Poisson's ratio, yield stress and so on, are firstly measured as for the design of structures using these multifunctional materials.   However, it is also necessary to evaluate the microscopic mechanical properties because the fracture might occur locally due to their heterogeneous structure.
        In the case of the fiber reinforced composite materials, the fracture behavior at the interface between fiber and matrix (or interlayer attaching with fiber) is one of the key issues for evaluating the mechanical properties of the composites.   Therefore, a fiber push-out test or a tensile test using a mini-composite with a single fiber has been conducted to examine the fracture behavior at the interface.[1-4]   Although the relationship between load and displacement can be dynamically measured in these tests, it is difficult to observe the fracture behavior directly. Many efforts concerning in-situ observation of fracture behavior in materials science field have been provided last decade, however it is hard to obtain the crack initiation site and the crack path by the fractography after testing.[5]
        Recently, Shibayama who is co-author of this paper has successfully developed a nano-mechanics in-situ transmission electron microscope (TEM) experimental apparatus as

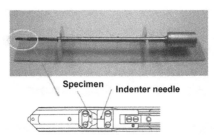

Figure 1.   Nano-mechanics in-situ TEM experimental apparatus.[6]

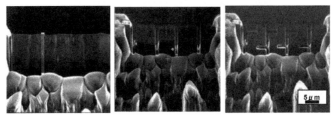

Figure 2.   Sequence of miniaturized DNS specimen preparation.

shown in Figure 1 to observe the fracture behavior of SiC/SiC composite and its joining directory in high voltage electron microscope (HVEM).[6, 7]   Because HVEM has superior advantages on penetration depth and 3D geometry of TEM specimen compared to a conventional TEM.   In this study, a new nano-mechanics in-situ TEM experimental apparatus was developed for measuring applied load and indentation depth (displacement) curve during the in-situ observation by using a prototype micro-electro-mechanical systems (MEMS) device.   By using this new apparatus, the fracture morphology of miniaturized double notched shear (DNS) specimen, which was prepared for TEM by focused iron beam (FIB) equipment, was examined. In addition, the finite element analyses with the interface element[8-11] were conducted in order to evaluate the fracture strength at the interface measured using this new apparatus.

EXPERIMENTAL TEST

Experimental Procedure
        The specimen used for DNS test was NITE (nano-powder infiltration and transient eutectic process) SiC/SiC composite.[12]   Since the composition of matrix is almost stoichiometric and its structure is high crystalline, it is rather easy to prepare the miniaturized DNS specimen.   After mirror polishing on the surface of NITE SiC/SiC composite by using a diamond slurry, the specimen was cut down to the small piece ($2 \times 3 \times 0.4$ mm^3) using a diamond cutting saw and then each side was polished by the diamond slurry again.   Finally, it was ground up to 100 mm thickness.   From this piece, a TEM thin film specimen was made using FIB equipment (30keV, Ga ion source).   Moreover, the TEM thin film region in the small piece of NITE SiC/SiC composite was also processed into a miniature sized DNS specimen by FIB. The shape of the miniaturized DNS specimen was compliant with the ratio of the length and the width of the small tab and the length of the notches and the distance between them in ASTM standards for the DNS testing method for ceramic composites,[13] except for its thickness.

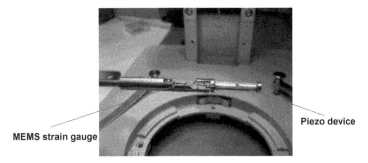

Figure 3.   A new nano-mechanics in-situ TEM experimental apparatus with MEMS device.

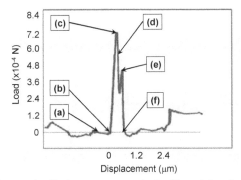

Figure 4.   Load – displacement curve obtained by miniaturized DNS test.

Therefore, the shear stress could be homogeneously distributed at the joining interface by the miniaturized DNS specimen.   Figure 2 shows scanning electron microscope (SEM) images of the DNS specimen preparation sequence.   A slight white line perpendicular to the notches between them is corresponding to the carbon (graphite) layer between SiC fiber and SiC matrix.

A new nano-mechanics in-situ TEM experimental apparatus consists of a piezo driven nano indenter and a MEMS to measure an applied load as shown in Figure 3.   A crack was introduced and propagated in this specimen using the new in-situ TEM apparatus in HVEM (Hitachi H-1300 operated by 1 MeV).   By using this apparatus, not only crack propagation but also applied load – indentation curve can be observed in sub micrometer scale during the test. Also, the results of in-situ observation were recorded by the high speed digital camera (Photron FASTCAM SA4).

Experimental Results
The applied load – indentation curve measured in the shear test using the miniaturized DNS specimen of NITE SiC/SiC composite was shown in Figure 4.   The series of TEM images extracted during in-situ observation on fracture behavior were summarized in Figure 5 where the extracted positions were plotted in Figure 4.   Black, gray and white granular contrasts are corresponding to each SiC grain in the fiber and matrix in the range of several ten nano meters. With increasing indentation depth (from Figures 5(a) to (c)), their Bragg diffraction contrasts changed gradually due to compression.   The carbon layer is highlighted by doted lines in Figure

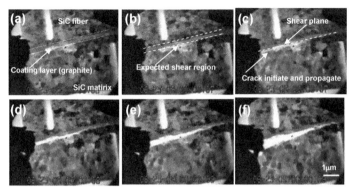

Figure 5.   TEM images of in-situ observation of fracture behavior during shear test
using miniaturized DNS specimen of NITE SiC/SiC composite.[6]

5(a).   Each TEM image in Figure 5 is the image (a) before testing, (b) start testing, (c) crack
initiation, (d) crack propagation, (e) increasing crack opening angle and (d) crack penetrated and
fractured, respectively.   Generally, the main crack is expected to propagate between the notch
roots in the conventional DNS test.[14, 15]   In this study, the crack started from the near-loading
point as shown in Figure 5.   However, the crack propagated straightly between the SiC matrix
and the coated carbon layer on the SiC fiber.   In addition, this crack propagation behavior has a
good agreement with the previous microstructural investigation about the crack initiation and
path by using the specimen after the fiber push-out test.   From the applied load – indentation
curve in Figure 4, the inter-laminar shear strength of NITE SiC/SiC composite in this study was
estimated as 2.8×10^3 MPa, which is about thirty times higher than the result obtained by the
conventional DNS test.[16]   In general, a failure of ceramics is strongly affected by the
distribution of defects and the number density of defects on the surface.   Figure 5 indicates that
this TEM specimen would have relatively lower density and distribution of defects and this
tendency would contribute the extremely high shear strength in this study.   Therefore, in order
to examine the reason of this high shear strength measured through the miniaturized DNS
specimen, the finite element analyses with the interface element were conducted.

NUMERICAL ANALYSIS

Model for Analysis
        The size of finite element model was 20 μm in length, 12 μm in width and 0.2 μm in
thickness according to the miniaturized DNS specimen used in the experiment as shown in
Figure 6.   The length and width of notch were 6 and 0.2 μm, respectively, and the distance
between the notches was 12 μm.   Although there was the coated carbon layer on the SiC fiber
in the experiment, the numerical model was simplified so that the carbon layer was avoided
because the crack did not propagate inside of the carbon layer.   In order to simulate the crack
propagation behavior, the interface elements were arranged along the interface between the SiC
fiber and matrix.   The total number of elements and nodes are 8120 and 8402, respectively.
        The mechanical properties of the SiC fiber and SiC matrix were assumed to be the same
since their microstructures seem to be almost the same according to TEM images as shown in
Figure 5.   Young's modulus and Poisson's ratio were set as 500 GPa and 0.3, respectively
because both the fiber and matrix were considered to have relatively low density and distribution

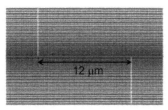

Figure 6. Finite element model of miniaturized DNS specimen.

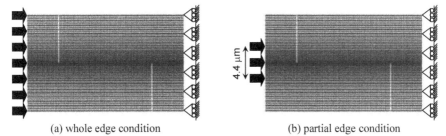

(a) whole edge condition   (b) partial edge condition

Figure 7. Mechanical boundary condition for analysis of miniaturized DNS test.

of defects and their mechanical properties are expected to have the higher values than those of SiC bulk.   Meanwhile, the yield stress was assumed to be 300 MPa or 1.5 GPa since the density of defects would largely affect the yield stress based on the theory of crystal plasticity.[17]
Two types of boundary condition were studied in this research as shown in Figure 7. The one models an ideal DNS test where the left edge of specimen was assumed to be deformed equally.   In the other boundary condition, a forced displacement was applied at a part of left edge as same as the experiment as shown in Figure 5.

Interface Element
Essentially, the interface element is the distributed nonlinear spring existing between surfaces forming the interface or the potential crack surfaces as shown by Figure 8.   The relation between the opening of the interface δ and the bonding stress σ is shown in Figure 9. When the opening δ is small, the bonding between two surfaces is maintained.   As the opening δ increases, the bonding stress σ increases till it becomes the maximum value σ_{cr}.   With further increase of δ, the bonding strength is rapidly lost and the surfaces are considered to be separated completely.   Such interaction between the surfaces can be described by the interface potential. There are rather wide choices for such potential. The authors employed the Lennard-Jones type potential because it explicitly involves the surface energy γ which is necessary to form new surfaces.   Thus, the surface potential per unit surface area ϕ can be defined by the following equation.

$$\phi(\delta_n, \delta_t) \equiv \phi_a(\delta_n, \delta_t) + \phi_b(\delta_n) \tag{1}$$

$$\phi_a(\delta_n, \delta_t) = 2\gamma \cdot \left\{ \left(\frac{r_0}{r_0 + \delta} \right)^{2N} - 2 \cdot \left(\frac{r_0}{r_0 + \delta} \right)^N \right\}, \quad \delta = \sqrt{\delta_n^2 + A \cdot \delta_t^2} \tag{2}$$

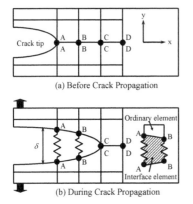

(a) Before Crack Propagation

(b) During Crack Propagation

Figure 8. Representation of crack growth using interface element.

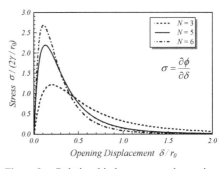

Figure 9. Relationship between crack opening displacement and bonding stress.

$$\phi_b(\delta_n) = \begin{cases} \dfrac{1}{2} \cdot K \cdot \delta_n^{\,2} & (\delta_n \le 0) \\ 0 & (\delta_n \ge 0) \end{cases} \tag{3}$$

Where, δ_n and δ_t are the opening and shear deformation at the interface, respectively. The constants γ, r_0, and N are the surface energy per unit area, the scale parameter and the shape parameter of the potential function. In order to prevent overlapping in the opening direction due to a numerical error in the computation, the second term in Eq. (1) was introduced and K was set to have a large value as a constant. Also, to model an interaction between the opening and the shear deformations, a constant value A was employed in Eq. (2). From the above equations, the maximum bonding stress, σ_{cr}, under only the opening deformation δ_n and the maximum shear stress, τ_{cr}, under only the shear deformation δ_t are calculated as follows.

$$\sigma_{cr} = \frac{4\gamma N}{r_0} \cdot \left\{ \left(\frac{N+1}{2N+1} \right)^{\frac{N+1}{N}} - \left(\frac{N+1}{2N+1} \right)^{\frac{2N+1}{N}} \right\} \tag{4}$$

$$\tau_{cr} = \frac{4\gamma N \sqrt{A}}{r_0} \cdot \left\{ \left(\frac{N+1}{2N+1} \right)^{\frac{N+1}{N}} - \left(\frac{N+1}{2N+1} \right)^{\frac{2N+1}{N}} \right\} \tag{5}$$

By arranging such interface elements along the crack propagation path as shown in Figure 8, the growth of the crack under the applied load can be analyzed in a natural manner. In this case, the determination on the crack growth based on the comparison between the driving force and the resistance as in the conventional methods is avoided. In the parameters involving above two equations, the shape parameter N and the interaction parameter between the opening and shear deformations were determined as 4 and 2.47 x 10^{-2}, respectively according to our previous research.[10, 11] As for the other two parameters, the surface energy γ and the scale parameter r_0 were assumed to 450 N/m and 0.4 μm, respectively, so that the maximum shear stress τ_{cr} becomes 2.8 x 10^3 MPa which was the measured value by using the miniaturized DNS specimen.

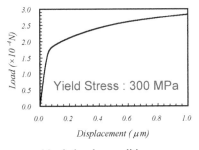

(a) whole edge condition

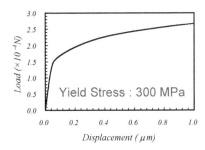

(b) partial edge condition

Figure 10.   Load – displacement curves computed by finite element analysis with interface element (yield stress of SiC = 300 MPa).

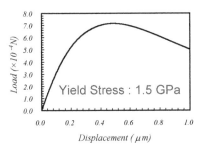

(a) whole edge condition

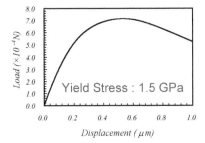

(b) partial edge condition

Figure 11.   Load – displacement curves computed by finite element analysis with interface element (yield stress of SiC = 1.5 GPa).

Numerical Results

The relationships between the displacement and the load calculated are summarized in Figures 10 and 11.   Also, von Mises stress distributions with 0.5 mm displacement are shown in Figures 12 and 13.   From these figures, it is found that the area for the forced displacement would not affect the load – displacement curves while von Mises distributions were slightly influenced.   Especially, in the case of a partial edge condition, the stress at a tip of the left notch becomes slightly larger in comparison with the results of a whole edge condition.   However, the crack propagation process was not affected, so that there are no differences in the load – displacement curves between these conditions.

On the other hand, from Figures 10 and 11, it is clearly revealed that the yield stress of SiC would have a great influence on the load – displacement curve.   In the case of lower yield stress, the load increased with increasing the displacement after achieving the yield point which was corresponding to the crack initiation.   While, in the case of higher yield stress, the load had a maximum value and then decreased with increasing the displacement.   This behavior has a fairly good agreement with the experimental result as shown in Figure 4.   In the experiment, the load drastically decreased after achieving the maximum point because the crack propagated suddenly after the initiation.   However, in this computation, it is difficult to demonstrate such a

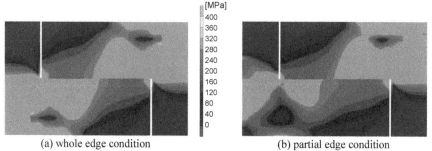

(a) whole edge condition   (b) partial edge condition

Figure 12. Von Mises stress distributions computed by finite element analysis with interface element at 0.5 μm displacement (yield stress of SiC = 300 MPa).

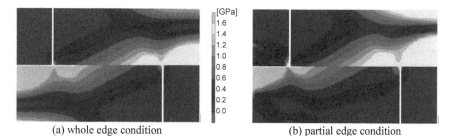

(a) whole edge condition   (b) partial edge condition

Figure 13. Von Mises stress distributions computed by finite element analysis with interface element at 0.5 μm displacement (yield stress of SiC = 1.5 GPa).

dynamic crack propagation behavior, since the calculations were conducted assuming a quasi-state behavior like a slow crack growth.[9] Nevertheless, the maximum load computed had a very good agreement with the experimental result and the displacement at the maximum load was also almost same as the experiment. Therefore, it can be concluded that this numerical analysis with a higher yield stress seems to demonstrate the experimental result of the miniaturized DNS specimen. Moreover, the reason of the extremely high shear strength in the experiment of the miniaturized DNS specimen is considered to be the higher yield stress of SiC fiber and matrix due to the lower density and distribution of defects.

CONCLUSIONS

As one of the in-situ experimental apparatuses, a new nano-mechanics in-situ TEM experimental apparatus was developed which is consisted of a piezo driven nano indenter and a MEMS to measure an applied load. By using this new apparatus, a compression test with a miniaturized DNS specimen of NITE SiC/SiC composite was conducted. Also, the finite element analysis with the interface element was conducted for examining the shear strength measured. The conclusions can be summarized as follows.

(1) The crack initiation and propagation at the interface between SiC matrix and carbon layer coated on SiC fiber are successfully observed with measuring the load – displacement curve.

(2)   The inter-laminar shear strength of NITE SiC/SiC composite was estimated as 2.8 x 10^3 MPa, which is about thirty times higher than the result obtained by the conventional DNS test.

(3)   By assuming a higher yield stress of SiC, the maximum load computed had a very good agreement with the experimental result and the displacement at the maximum load was also almost same as the experiment.

(4)   The reason of the extremely high shear strength of the miniaturized DNS specimen seems to be the higher yield stress of SiC due to the lower density and distribution of defects.

REFERENCES

[1] H. Serizawa, A. Kohyama, K. Watanabe, T. Kishi and S. Sato, Elastic FEM Analysis of Fiber Push-Out Test for C/C Composites, *Materials Transactions, JIM*, **37** (3), 409-413 (1996).

[2] S. Sato, K. Watanabe, H. Serizawa, K. Hamada and A. Kohyama, Effect of Heat Treatment Temperature on Interfacial Mechanical Properties of C/C Composites by means of Micro-Indentation Test, *Proceeding of International Conference on Microstructures and Functions of Materials*, 157-160 (1996).

[3] T. Hinoki, W. Zhang, A. Kohyama, S. Sato, and T. Noda, Effect of Fiber Coating on Interfacial Shear Strength of SiC/SiC by Nano-Indentation Technique, *Journal of Nuclear Materials*, **258-263**, 1567-1571 (1998).

[4] R.J. Kerans, F. Rebillat and J. Lamon, Fiber-Matrix Interface Properties of Single-Fiber Microcomposites as Measured by Fiber Pushin Tests, *Journal of the American Ceramic Society*, **80** (2), 506-508 (1997).

[5] M.A. Wall and U. Dahmen, An In Situ Nanoindentation Specimen Holder for A High Voltage Transmission Electron Microscope, *Microscopy Research and Technique*, **42**, 248-254 (1998).

[6] T. Shibayama, T. Ogitsu, S. Yatsu and S. Watanabe, In-situ Observation of Crack Propagation in SiC/SiC by HVEM, *Abstract of 13rd International Conference on Fusion Reactor Materials*, 3115 (2007).

[7] T. Shibayama, G. Matsuo, K. Hamada, S. Watanabe and H. Kishimoto, In-situ Observation of Fracture Behavior on Nano Structure in NITE SiC/SiC Composite by HVEM, *IOP Conference Series: Materials Science and Engineering*, **18**, 162013 (2011).

[8] Z. Q. Wu, H. Serizawa and H. Murakawa, New Computer Simulation Method for Evaluation of Crack Growth Using Lennard-Jones Type Potential Function, *Key Engineering Materials*, **166**, 25-32 (1999).

[9] H. Murakawa, H. Serizawa and Z. Q. Wu, Computational Analysis of Crack Growth in Composite Materials Using Lennard-Jones Type Potential Function, *Ceramic Engineering and Science Proceedings*, **20** [3], 309-316 (1999).

[10] H. Serizawa, H. Murakawa, M. Singh and C.A. Lewinsohn, Finite Element Analysis of Ceramic Composite Joints by Using a New Interface Potential, *High Temperature Ceramic Matrix Composites 5*, 451-456 (2004).

[11] H. Serizawa, K. Katayama, C. A. Lewinsohn, M. Singh and H. Murakawa, Numerical Analysis of Mechanical Test Methods for Evaluating Shear Strength of Joint By Using Interface Element, *Key Engineering Materials*, **345-346**, 1489-1492 (2007).

[12] Y. Katoh, A. Kohyama, T.Nozawa, M. Sato, SiC/SiC Composites through Transient Eutectic-phase Route for Fusion Applications, *Journal of Nuclear Materials*, **329-333**, 587-591 (2004).

[13] E. Lara-Curzio and M.K. Ferber, Shear Strength of Continuous Fiber Ceramic Composites, *Thermal and Mechanical Test Methods and Behavior of Continuous-Fiber Ceramic Composites*, ATSM STP1309, 31-48 (1997).

[14] T. Hinoki, L.L. Snead, Y. katoh, A Kohyama, R. Shinavski, The Effect of Neutron-Irradiation on the Shear Properties of SiC/SiC Composites with Varied Interface, *Journal of Nuclear*

Materials, **283-287**, 376-379 (2000).

[15] T. Hinoki, W. Yang, T. Nozawa, T. Shibayama, Y. Katoh, A. Kohyama, Improvement of Mechanical Properties of SiC/SiC Composites by Various Surface Treatments of Fibers, *Journal of Nuclear Materials*, **289**, 23-29 (2001).

[16] T. Hinoki, Development of Evaluation and Application Techniques of SiC/SiC Composites for Fusion Reactors, *Journal of Plasma and Fusion Research*, **80**, 31-35 (2004).

[17] J. Lubliner, *Plasticity Theory*, Pearson Education Inc., 2006.

COMMERCIALISING UNIVERSITY RESEARCH: THOUGHTS ON THE CHALLENGES BASED ON EXPERIENCE GAINED IN THE FIELD OF CERAMIC PROCESSING IN THE UK

Jon Binner
Department of Materials, School of Aeronautical, Automotive, Chemical and Materials Engineering, Loughborough University, Loughborough, UK

ABSTRACT
    Transferring technology from University to industry is rarely a simple matter and there are many challenges that need to be overcome. This paper outlines some of the lessons learnt, both positive and negative, during a UK-based career spanning 3 decades in the field of ceramic processing that has seen several different technologies transferred into industry, via both spin out and licensing routes – and a number of other technologies fail to be transferred. The paper will also outline how properly designed Government funding schemes can be of tremendous assistance in the process of crossing the so-called 'Valley of Death'.

INTRODUCTION
    It is strangely disturbing to realise that in an academic career that began in 1984, the sum total of research that has made it into industry is limited to the creation of a spin out company for a ceramic-based product for measuring soil matric potential; licensing technology for producing ceramic foams and a license agreement for producing nanostructured zirconia components. All other developments worked on have failed to make it into industry; although I do like to add the rider 'yet'! Nevertheless, this seems scant return for what is now over 28 years of research and development involving a total of ~£11M (~$17.8M) of funding and 34 PhD students and 27 postdoctoral research projects. Since failure can be as educational as success, I believe that it is possible to draw some conclusions from the various attempts to get ideas researched, developed and commercialised. So, by virtue of a few case studies, I should like to present these ideas for consideration by others in the hope that they will be helpful and save others from making some of the (many) mistakes that I have made.

CASE STUDY 1: SOIL MATRIC POTENTIAL
    Soil matric potential is a measure of the availability of moisture in soil. In an unsaturated soil, the matric potential results from capillarity and adhesion forces. Plants must overcome the energy of the matric potential to extract water from the soil. The device developed was intended to overcome the shortcomings observed with existing techniques. It consisted of two primary components:
i)   a porous piece of ceramic in which the porosity was in the form of straight channels of uniform cross section and known diameter.
ii) a very low power, microwave-based detector that would determine the moisture content of the ceramic from the real part of the permittivity.
    The principle of operation was very simple. Capillary channels would fill with water or empty depending on the availability of moisture in the ground. By burying a piece of ceramic containing parallel channels of a given size in the ground, the channels would fill or empty as the moisture level in the ground changed. The microwave-based detector then determined the water content of the ceramic and from this information a small microprocessor decided whether the ground needed irrigating or not.
    In reality, there were various levels of complexity for the instrument. Consider the following three examples, though a number of other formats were envisaged:

i) A single block with a single channel size. This device would act as a simple yes/no monitor. If the block was empty of water, the ground was drier than a certain critical level and so irrigation would switch on for a predetermined time. The block could be designed to fit the wetness requirements of a specific crop in a specific soil.

ii) Two blocks, each with one, different, channel size. Irrigation was controlled by the microprocessor, which aimed to keep one block dry and the other wet, thus maintaining the desired water content in the soil. Again the device was based on a simple yes/no criteria.

iii) A more sophisticated device was a single block with a controlled range of channel sizes. The amount of moisture in the block was measured, rather than simply whether the block was dry or not. From this information, and knowing the channel characteristics of the block, a numerical value could be placed on the soil water potential. Such an instrument would be aimed at the research market rather than the farming community.

The ceramic component was manufactured using fairly standard composite forming technology[1], figure 1. Carbon fibres were passed through an alumina/silica slurry and wound onto a drum an appropriate distance apart to ensure that there was sufficient matrix between them. They were then cut and laid flat and pressed to form a green body that was subsequently dried and then sintered. During sintering the carbon fibres burned out leaving uniform, aligned channels. Provided that there was sufficient material between the separate channels the latter remained unconnected. The channel width was controlled either by the initial oxidation of the fibres[2] and/or by back infiltrating the channels using an aluminosilicate sol gel[1].

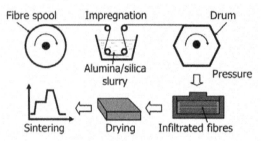

Figure 1: Manufacturing route for ceramics containing uniform channels of controlled diameter.

The particular advantages of the device were its simplicity and the lack of hysteresis, moving parts, or being affected by temperature or the salt content in the soil, e.g. from the use of fertilisers. Although originally intended for use in the agricultural industry, it was considered from the start that it would also have applications in the construction and hydroengineering industries.

In 1992 a spin out company, Ceratronics, was created by Dr Tom Cross, a colleague in the Dept of Electrical & Electronic Engineering, and me. The company ran for about 4 years, but we weren't happy with the management arrangements imposed on us by our University and eventually decided to let the company die. I believe that management of spin-outs is now handled much better at the University where we were at the time. Nevertheless, we both learned a lot. Not least, the need to be able to control own destiny, for better or worse. With hindsight, we probably also created the company at least 2 – 3 years too soon. Whilst we had the basic ideas worked out we were far short of having an even remotely working prototype. Things always take longer than anticipated and there is a clear need to have a reasonably secure source funding for several years. My memory is of spending more time chasing the next bit of funding than I spent working on the technical issues. In this respect, we would have made much faster progress if we had had a team, rather than a lone researcher, but funding was very limited. In addition, we made a fatal mistake in not clearly identifying the best market to attack. It rapidly became clear that water was just not expensive enough in the UK to justify the extensive development costs. Although there was significant interest from organisations in South Africa, the Rand was too weak and the potential backers could not afford the costs; they would have paid roughly ten times more for us to do the development in the UK compared to someone else doing it in S Africa. We should have been willing either to license the technology to them and sit back and wait for the royalties (but who wants to give away what they see as their brainchild?) or, seeing the international potential, we should have tried other countries where water is more scare than in the UK and whose currencies were strong, e.g. Australia or Israel. However, by the end, we realised that our biggest mistake of all was not attacking the greenhouse market. Here, watering is always required and a fully automated system that delivered just the right amount could have significantly reduced water bills, even in the UK where water was, and still is, a relatively cheap commodity.

CASE STUDY 2: CERAMIC FOAMS

At roughly the same time that I was involved with Ceratronics, I was also working with a company called Dytech Corporation Ltd in the UK on a new manufacturing route for foam ceramics based on gel casting. The latter process was originally developed by ORNL[3]. Gel casting employs an organic monomer that is polymerised to cause the in-situ gelation of a foamed aqueous ceramic slurry. The primary advantage of the process is its inherent flexibility. The foams could be near net shape manufactured in a variety of shapes and sizes, figure 2, and after production were simply dried and fired. They could be machined easily in both the green & fired state and hole drilling, turning, slitting, etc., were all possible. In addition, the porosity and pore size could be varied to suit the application and a wide range of ceramic materials could be foamed. The criterion was the ability to produce a stable ceramic suspension. Foams were produced in alumina, partially and fully stabilised zirconia, mullite, cordierite, hydroxyapatite and barium titanate, as well as silicon carbide and aluminium nitride. Densities ranged from 5% to 40% of theoretical. Potential applications were diverse and ranged from furnace linings, high temperature kiln furniture and crucibles, to catalyst supports, hot gas filtration and artificial bone[4].

Figure 2: Examples of the variety of shapes that could be produced from the ceramic foams.

Figure 3 shows the process flow chart; it can be seen that it is quite complex and some of the chemicals used required an absence of oxygen and so glove box technology was required. The latter complicated matters significantly, though it was still possible to scale up and commercialise the process whilst keeping it economically viable[5]. Figure 4 shows the structure of the foams produced. For one particular application it was possible to apply a dense, impermeable coating to one or more sides, eliminating the inherent permeability of the foams whilst maintaining the other properties. More recently, it has been extensively demonstrated that it is possible to infiltrate the foams with a range of aluminium-based molten alloys and polymers to make interpenetrating composites that, themselves, have a wide range of desirable properties[6,7].

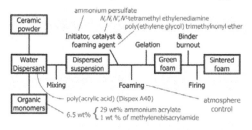

Figure 3: Process flow chart for producing ceramic foams by gel casting.

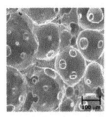
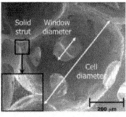

Figure 4: Structure of the gel cast ceramic foams.

This time the research was undertaken by a reasonable size of research group. Over a period of about 5 years it involved two EPSRC PhD CASE Awards[a], two 100% industry funded PhD projects and a Teaching Company Scheme[b] to transfer the technology to Dytech. A number of patents were taken out to cover the technology. As well as Dytech exploiting the technology it seems that a couple of employees left and created their own, separate companies that also exploited it in different niches. After several years, the technology was transferred to Dyson Thermal Technology (Dyson TT), a sister company within the overall Dyson Refractories organisation and they continued to support further developmental work at the University. It is believed that Dyson TT did not survive the recent recession and it seems that one of the other two companies also did not make it. However, Ceramisys has survived and continues to produce hydroxyapatite ceramic foams for use in applications such as bone grafts. More information can be found at: http://www.ceramisys.com.

Positive Lessons Learnt

- Even quite complex technology can be transferred successfully to industry.
- Need to consider manufacturing costs from the outset.
- Teaching Company Scheme, TCS (now Knowledge Transfer Partnerships, KTP) was/is an excellent scheme for technology transfer and provides good training for the individual.
- Using the individual who did the original research makes things much easier.
- Having a team at the University, not a single PhD student, results in much faster progress (once it is established there is something worth pursuing).
- You learn a lot when transferring technology!

Negative Lessons Learnt

- Things always take longer than you anticipate.
- Need to have a very clear idea of the markets and the order in which to attack them.
- Launching a product too early can be quite damaging.
- 'Mollycoddling' the researchers is not appreciated by the company if they are subsequently employed.
- Trust and openness are needed in both directions.
- An eye needs to be kept on the University Research Office to ensure that they fully understand the ramifications of all the developments.
- Avoid company politics at all costs!
- Think before becoming too dependent on the company providing samples for future research.

Table 1: Positive and negative lessons learnt from licensing the ceramic foam technology.

---

[a] The Engineering & Physical Sciences Research Council (EPSRC) funds research at universities over the approximate TRL range of 1 to 3 or 4. CASE Awards are PhD programmes jointly funded by EPSRC and an industrial company.

[b] Teaching Company Scheme, now called the Knowledge Transfer Partnerships (KTP), a TSB-funded, technology transfer programme in which one or more recent graduates work for 12 – 36 months in a company whilst being supported by a university.

The extensive research, development and technology transfer work undertaken over more than a decade in the field of ceramic foams has led to a much greater understanding of the whole field of technology transfer. The bullet points in Table 1 above summarise both the positive and negative lessons learnt. Most of them are extremely obvious and really shouldn't have needed learning, but the issues underpinning them rarely present themselves in an obvious way at the time. For example, when a global aerospace company asked for some of the alumina foam to evaluate it for a particular, but confidential, application, there was a lot of pressure to provide the foam immediately without asking any questions at all. The foam evaluated failed completely, causing the aerospace company to lose interest. Many years later the application was revealed during other discussions with the aerospace company and it was realised that an alumina foam was probably the worst possible one that could have been chosen! Open communications under a non-disclosure agreement could have led to a new product for Dytech and a problem solved for the aerospace company.

CASE STUDY 3: NANOSTRUCTURED CERAMICS

A wide range of different bulk nanostructured ceramics have been and are being developed at Loughborough University, these include alumina, barium titanate, hafnium diboride and carbide, iron oxide, yttrium aluminium garnet, zinc oxide, zirconia and zirconia toughened alumina. The intended applications range from electronics to armour, valves to dental ceramics, solid oxide fuel cells to thermal protection systems. Each material has a different source of funding, from Government-based to industrial, and is at a different stage of the development process. The material that has been taken the furthest is yttria partially stabilised zirconia; the underpinning technology has now been licensed by MEL Chemicals in the UK. This development has also seen the largest number of researchers working on it, with no less than three PhD students and five postdoctoral research associates contributing over the decade or so that the research has been underway.

In total, three patents have been developed to date associated with the processing and properties of the nanozirconia. Since the starting material is dilute nanopowders suspensions produced by MEL Chemicals itself, the first patent is associated with their concentration to the solids contents required for either slip casting or granulation without a concomitant increase in viscosity. It is the latter that is so difficult. As suspensions become more concentrated the particles are forced closer together and van der Waals forces become increasingly dominant; with nanoparticles the effects are particularly dramatic and the viscosity increases very sharply with increasing solids content. However, it has been demonstrated[8] that a combination of the right surfactant and the use of ultrasound to break up any agglomerates that form yields the ability to concentrate the nanosuspensions whilst keeping the viscosity low. The resultant suspensions can be both slip cast and, following a very significant amount of developmental work, die pressed. For slip casting, the key step is associated with the resultant drying of the body; the very narrow channels between the nanoparticles make drying without cracking difficult, the use of humidity drying does help however[9].

Industry's favoured green forming route, however, is dry forming; it is both much faster and much cheaper than any of the wet forming routes. The problem is that nanopowders simply do not flow due to their massive surface areas, and hence surface energies. Granulation is therefore required to create the necessary flowability. Spray drying was found to yield incredibly hard granules that did not crush even at die pressing pressures of 500 MPa and in general, the ceramics industry prefers to use pressures of no more than about 250 MPa, ideally 200 MPa or less. Whilst some excellent work has been done in Germany based on spray drying hollow granules[10], our approach was to turn to spray freeze drying. The use of cryogenic, rather than elevated temperatures, very significantly reducing the strength of the granules[11]. The combination of spray freeze drying and addition of foaming agents, which has been patented, has allowed green bodies to be formed using industrial presses; densities were around 55% of theoretical and the bodies were capable of being sintered to full density whilst retaining a nanostructure.

Figure 5 shows a comparison of the microstructure of the nanozirconia at the same magnification as a more conventional material produced from Tosoh zirconia powder. Both are

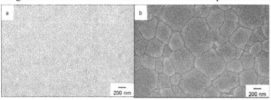

>99.5% of theoretical density. The Tosoh powder, sintered at 1500°C for 2 hrs, yielded a mean grain size of ~0.5 μm whilst the nanoceramic has a mean grain size of ~65 nm. The latter can be controlled by the sintering conditions used; two-stage sintering using microwave hybrid heating, in which both microwaves and conventional radiant heating are combined, yields the finest grain sizes[12]. The

Figure 5: Comparison of nano and submicron 3YSZ ceramics at the same magnification showing a) a nanostructured body and b) a submicron ceramic made from Tosoh powder, see text for details.

optimum sintering conditions for 3 mol% yttria doped zirconia were 6 s at 1150°C for T1 followed by 3 hours at 1050°C for T2 for slip cast bodies. However, when the yttria content was reduced to 1.5 mol%, the two-step sintering cycle needed modification; the lower the yttria level the less demanding the sintering conditions required.

The strength of the nanoceramics has been found to be very similar to that of conventional submicron ceramics, ~1 GPa, although the fracture mechanism is different; the nanostructured ceramics all fail via an intergranular fracture mode whilst failure was transgranular for the submicron ceramics.. Two toughness measurement approaches have been used, indentation and surface crack in flexure. The results indicate that the nano 1.5YSZ ceramics may be best viewed as crack, or damage, initiation resistant rather than crack propagation resistant; indentation toughness measurements as high as 14.5 MPa m$^{1/2}$ were observed. The wear mechanism of nanozirconia has been observed to be different compared to that in conventional, submicron YSZ and the wear rates to be lower, particularly under wet conditions. In addition, and potentially most usefully, the nano 3YSZ ceramics appear to be completely immune to hydrothermal ageing, even when not fully dense, for up to 2 weeks at 245°C & 7 bar; conditions that see a conventional, commercial submicron ceramic disintegrate completely within 1 hour. This latter property has seen the 3rd patent application filed.

Just as for the other case studies, the work has led to a number of conclusions being drawn relating to the process of technology transfer and commercialisation, table 2. Of these, most are again fairly obvious, but perhaps one or two points could do with further expansion. The sequence of research grants, and the wonderful technology transfer nature of the EPSRC Follow On Fund and Collaboration Fund schemes, really helped the team to accelerate the process of getting their ideas into industry. Unless something unforeseen happens, the technology will be commercialised within around a decade of the start of the research; something that is really quite rare. The team considers itself very lucky, however. Each time, just when it was needed, there was a Call for Proposals in just the right area and in just the right format. What is needed, is to remove some of the luck from this process. Government agencies need to talk to each other and ensure that there is a small amount of funding reserved by, in this case, EPSRC and the TSB[c], that can only be applied for by researchers who have already been successful and are developing

---

[c] The Technology Strategy Board (TSB) funds collaborative research between universities and industry over the approximate TRL range of 3 or 4 to 5 or 6.

their ideas steadily closer towards commercialisation. The funding must certainly remain competitive, there must be no diminution of the principle of funding excellence, but such a fund would remove the element of chance that there is a Call in an appropriate area; something that is currently quite restrictive for the TSB in the UK in particular. It is believed that this development of nanostructured ceramics is an achievement that exemplifies how the availability of successive and appropriate public funding initiatives can lead to effective innovation. Removing the element of chance that has been involved can only improve this process further.

Positive Lessons Learnt

- Having a team at the University, not a single PhD student, results in much faster progress.
- Having talented researchers makes life much easier.
- When you gain the trust of the company, progress is faster.
- The EPSRC Follow On Fund and Collaboration Fund schemes are / were excellent technology transfer mechanisms.
- Property patents are much more valuable than process patents.
- You need luck to get funding just when you need it.
- You need a LOT of luck to tie EPSRC and TSB funding together.
- Taking the time & effort needed to consider commercialisation issues and create market reports is beneficial.

Negative Lessons Learnt

- Need to consider manufacturing costs from the outset.
- Government funding needs joined up thinking and to acknowledge the so-called 'Valley of Death'.

Table 2: Positive and negative lessons learnt from licensing the nanozirconia technology.

The other comment that I would make on this issue is to report a conversation that I had over dinner one day about a year ago with a very senior individual in the UK Government

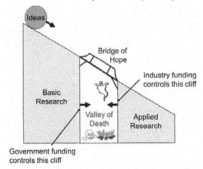

Figure 6: The so-called 'Valley of Death' for new inventions.

funding hierarchy that I happened to be sitting next to. I outlined the idea above of linked funding to lead research by the hand through to commercialisation with a view to bridging the so-called 'Valley of Death', figure 6. Whilst academic researchers and industrialists seem to be fully conversant with this idea I was very surprised to find its very existence being denied by my Government colleague. When I pressed the point, which was not well received, I got the response that, since there was felt to be nothing that the Government could do about it, the best approach was to simply ignore the issue and deny its existence. I did not get the impression that there was anything remotely teasing about this response and the subject was immediately changed.

LESSONS LEARNED FROM TECHNOLOGIES NOT TRANSFERRED

I believe that there are a number of reasons why many projects have failed to be commercialised, but I have, crudely, divided these into three fundamental groups, viz. poor timing, the wrong process route and the lack of a supply chain. Each of these is looked at below in turn.

a) Poor Timing

Three are three projects that I have put into this category; the first two involved multiple researchers working for several years, whilst the third involved just a single postdoctoral researcher and three years.

i)   Microwave enhanced chemical vapour infiltration

This was a a research programme that was initially funded by an excellent, though now deceased, research funding programme run by EPSRC in the UK. Known as Realising Our Potential Awards (ROPA), the idea was that academics that met a minimum criterion in terms of success in getting EPSRC funding could submit entirely blue sky ideas for possible funding. There was full peer review so it was no easy route to funding, but the existence of the criterion reduced the numbers applying and hence increased the odds of success. The idea involved in this case was to utilise the fact that when bodies are heated using microwaves they often develop an 'inverse temperature profile', i.e. the centre of the body becomes hotter than the surface since power is only deposited into the ceramic, the surrounding air remains cooler than the body[13]. The magnitude of the profile had been found to be dependent upon many factors including power level, electric field intensity and material properties such as dielectric loss, and thermal and electrical conductivity.

Chemical vapour infiltration (CVI) is used for producing continuous fibre reinforced ceramic matrix composites (CMCs). In the process a preform, such as a porous array of fibres, is heated in an atmosphere of gaseous reagents causing the latter to decompose and deposit a solid coating onto the fibre surfaces. Eventually the deposited material forms a dense matrix around the fibres. There are a number of advantages including processing temperatures significantly below those required for sintering, near net shape capability, and the ability to control microstructures. The primary difficulty with CVI technology is the tendency to deposit preferentially near the surface of the preform. This leads to pore blockage, or crusting, which inhibits densification and requires intermittent machining operations to remove the crust. The net effect is a non-uniform composite density and/or very long processing times, often of the order of weeks for large, complex components. This means that the CVI process is restricted to high value components such as those used in the aerospace industry. The problems stem from the heating methods used to heat the preform, which rely on surface heating and conduction of heat to the interior. What is required is a cool surface with a hot interior, exactly what microwave heating can achieve.

The work itself was successful and it was clearly demonstrated that microwave heating achieved the desired temperature profile; samples densified rapidly from the inside-out. Subsequent work funded by a UK Government military agency led to a degree of optimisation of the process; just 10 h of heating could yield ~50 mm diameter samples that were effectively fully dense in the middle and with an average density overall of ~75% of theoretical[14]. Conventional heating was required to densify the surface layers of the composites, however a hybrid heating system that was developed and constructed was moved to the military sponsor just weeks before it was reorganised and, effectively, broken up as an organisation. The equipment was, sadly, never seen again. Although it proved impossible at the time to get funding to construct a second facility, it is hoped that, after about a decade since the original work finished, we can restart it with the focus being on UHTC composites rather than the original SiC$_f$/SiC composites thanks to the receipt of a major grant of £4.2M (US$6.75) from the EPSRC under a scheme known as a Programme Grant. Led by me, the grant also includes Professors Bill Lee and Mike Finnis from Imperial College and Professor Mike Reece at Queen Mary University, both in London. The grant is to look at Material Systems for Extreme Environments and will include work on a wide range of UHTC materials for applications as diverse as hypersonic flight to nuclear fission and fusion.

ii) Nano zinc oxide

As was seen earlier, the work on nano zirconia has been licensed and should be commercialised during 2013. However, earlier industrially-sponsored work on nano zinc oxide, nano ZnO, was very successful and led to the production of varistor-grade materials that out-performed existing ceramics based on submicron technology. Commercially screen printed multilayer varistors (MLVs) and integrated passive devices (IPDs) were produced using a varistor grade, nano ZnO ink and then microwave sintered. The results demonstrated that the Nominal Voltage (the voltage measured, V1mA, corresponding to a 1 mA current through the varistor), the Clamp Ratio (the ability of the varistor, expressed by CR = Vc / V1mA, where Vc is the voltage measured for a given peak current Ic), the Leakage (the maximum current at which the varistor can be operated without getting significantly heated) and the Variance (the square root of the standard deviation of the Nominal Voltage values) were all improved, even after scaling up to 12,000 components per batch[15]. Processing was also an order of magnitude faster. Unfortunately, just as the results were achieved for the scaled up batches of materials the sponsor, an Irish multilayer device company, was bought by another organisation and production moved to China. The new owners saw little value in the technology developed, apparently believing that sufficient savings would be made by virtue of the new location of the plant. Once again, promising technology failed to be commercialised due to poor timing that was entirely outside of our control.

iii) Rotary moulding

A third example of where the timing was simply wrong, was the development of a process based on rotary moulding. This approach sought to produce a new route to the fast and economic production of hollow ceramic components. It combined in-situ coagulation moulding (see below) with a modified version of the technique of rotary moulding, the latter being adapted from the polymer industry. Colleagues from Queens University Belfast, in Northern Ireland, with expertise in the rotary moulding of polymers played a vital role. Following substantial optimisation work, it was found that a two-speed approach to multi-axial rotation was the most successful; medium sized cream jugs could be produced in just 7 minutes. With respect to mould materials, the porous resin normally used for pressure casting of sanitary ware was found to be the best option, though since this is quite expensive conventional plaster-of-paris moulds were found to be a suitable material to enable companies, particularly SMEs, to become familiar with the technology whilst avoiding high costs for trials. The processed articles could be successfully fired and glazed using gas-fired kilns. The major advantages of the process included the ability to precisely calculate the amount of ceramic slip required, eliminating either slip wastage or the need to pour used slip back into the virgin material as currently happens with slip casting. In addition, since the precursor suspension required a very high solids content, the time and energy required to dry the green product and associated moulds was considerably reduced[16].

The problem was that at the time when it was proposed to do the work, the only companies willing to sponsor it were in the traditional, clay-based sector and they were suffering very significant competition from overseas (which is why they funded the work; they were desperate). The whole development, from start to finish, took only 3 years and one postdoctoral researcher. Nevertheless, the results came too late and the primary sponsor was forced to contract and abandon the project just weeks before we were ready. A UK patent was filed, but, although there was a brief resurgence of interest by an advanced ceramics company a few years later when it was realised that hollow ceramic spheres could be produced for which they seem to have had a potential use, the work was never commercialised. One of my most abiding memories of this

R&D project was of regular visits in the last year of the project to the factory floor of the primary sponsor where a prototype was being constructed. The attitude of the workers was very difficult to handle; they knew that if the equipment was completed and worked as planned roughly half of them would probably be made redundant. However, if it didn't, then probably all of them would be; as I believe sadly proved to be the case.

b) Wrong Process Route – In-situ Coagulation Moulding

Interestingly, this process development came about as a result of a chat over a pint a beer with a colleague at Nottingham University, where I was working at the time, who was in the Food Science Department. He explained to me how chocolate was made by the coagulation of fats containing cocoa particles and I realised that it should be possible to produce green ceramic bodies by the same principle. Six years and two PhD students later, we had developed a fast, near-net shape route for the production of advanced ceramic components that used carboxylic acid derivatives as coagulants for electrosterically dispersed, high solids content ceramic suspensions[17]. The time dependent *in-situ* hydrolysis of the coagulant D-gulonic-γ-lactone (one of very many we investigated[18]) progressively destabilised the suspension to form a viscoelastic solid within which the homogeneity of the initial dispersion was maintained. Constraining this hydrolysis reaction within a non-porous mould led to the formation of green bodies that could be quite complex in shape, figure 7. After drying, the ceramic components could be sintered without a special debinding operation since only a small amount of organic additive, less than 1 wt% of the total batch mass, was needed. Since the forming process took place without pressure and at

Figure 7: An illustration of some of the shapes that could be formed with in-situ coagulation moulding.

temperatures around ambient conditions, inexpensive moulds and tools could be used. As applied to α-alumina, this new forming technology produced components that possessed high sintered densities of ~99 % theoretical, uniform microstructures, high mechanical strengths and reasonable Weibull modulus[17]. The only problem was that, unfortunately, no company was interested in the concept! It took me far too long into my career to work out that, generically, industry is far more interested in dry green forming routes than wet green forming routes. Wet routes are much slower and, due to the drying required, more energy intensive. Therefore, I have observed that they tend to be used only when other routes are not practical, often due to the complexity of the component shape. This is a considerable shame given that many, many more academics world-wide work on wet forming routes compared to dry forming routes! As indicated in the example above, there was some interest from the clay-based traditional ceramic sector, which is far more accepting of wet forming routes, when the in-situ coagulation moulding route was combined with rotary moulding, though, as we have seen, this interest was more out of desperation than anything and had been left far too late when the vultures were already actually landing, rather than merely circling. There are many wet forming routes, developed in other laboratories around the world, that have met similar fates with little or no commercialisation even after a decade or more of a working system being developed.

c) Lack Of Supply Chain – Microwave Sintering

This final entry into the pantheon of ideas that have, so far, failed to be commercialised, is probably the most controversial. I first began working on microwave sintering as long ago as 1985 after heating a bowl of canned chilli to eat one evening whilst watching television. I suddenly realised how quickly I had been able to get my meal ready (microwaves were not common in the UK in the mid-1980s) and wondered what could be done with, what to me was a brand new technology, with respect to ceramic processing. Sadly, within 24 hrs my illusions that I must be the first to think of this new idea were removed as I realised that others had been at work on the same theme for at least two decades. Nevertheless, over the quarter of a century I managed to establish a small reputation in the field, working not just on microwave sintering but many other applications including joining, slip casting, chemical vapour infiltration, drying, debinding, amongst others[19]. There is no doubt in my mind that 'pure' microwave sintering has very limited applications, firstly because many ceramics do not absorb microwaves sufficiently but mainly it is quite difficult to achieve accurate temperature control. Microwaves are also not usually found to be a cheap source of energy. However, the concept of hybrid sintering, or in the USA, microwave assist technology (MAT), in which both microwaves and conventional heat sources are combined, is one that I firmly believe does have a vibrant future. The concept was developed in the UK during the 1990s[20] at EA Technology, which subsequently became known as C-Tech Innovations Ltd. Products investigated ranged from advanced ceramics to bricks, sanitary ware to tableware, and included abrasives, pigments and refractories. Almost all products could be produced at twice the speed with equivalent or better quality and the results of cost benefit analyses showed that the payback period was typically less than 3 years[21]. A more recent analysis performed for a brick company in 2006 showed a payback of only 13 months, partly because of an increase in gas prices. The design of hybrid kilns also has advanced, leading to decreased kiln costs and encouraging more companies to investigate the potential for their products. One recent example, the firing of SiC sleeves for Blasch Precision Ceramics by Ceralink, the exclusive agents for MAT technology in the USA, has revealed that the temperature uniformity achievable in the part permitted five time faster heating rates to be utilised, resulting in 500% savings in time and an estimated 41% saving in energy costs over the existing process[21]. As far as I can tell, Ceralink is doing good business in the USA, but this technology should be world-wide by now. My suspicion is that the supply chain is simply not strong enough. There needs to be more activity outside of the USA, either by Ceralink being licensed to operate on a wider basis or by other companies becoming more active.

LICENSING VERSUS SPIN OUT

This paper has looked at a number of 'case studies' to draw conclusions about the difficulty of translating academic research into industry. It is time now to start to draw some of the ideas together and a good place to begin is to consider the alternatives of licensing versus spin out. I have tried to consider the concept of licensing from both an academic and an industrial perspective, though clearly the latter is an academic's view of what it must seem like to an industrialist!

a) Licensing (academic perspective)

There is no doubt that it is best to involve the industry partner as soon as possible. There is always a temptation to get the research fully developed and THEN to approach industry, but it is so easy to pursue the wrong idea or the right idea in the wrong way. It can be very painful to watch years of hard slog being swept away as the industrialists narrow in on the aspect that they think that they can exploit. I would also advise listening to what the industrialists say; they are the experts on commercialisation after all (though there have been many times when I have seriously wondered about this!). The point is, if they fail to make money, they go out of business and jobs are lost. Whilst this is definitely not a guarantee that all decisions made are the right ones, or companies would go out of business far less often than they do, it does mean that mistakes are generally learnt from. I would also strongly advise being honest and open with the industrialists. If something isn't working as expected, tell them. There is little that will upset an industrialist more than to find out that there was a nagging problem with the process all along that you, the academic, knew about but didn't disclose. Trust takes so long to build up but is such an easy thing to lose. I would also caution that you should expect quite unreasonable demands on your time (to which you will have to agree with good grace). Industrialists never seem to understand that research is only part of our job and that we also have to juggle the demands of teaching and administration and that there are parts of the year where research simply has to be put on the back burner for a week or three. My penultimate point is that if you chose to go the licensing route, you should never expect to become rich. The lawyers will argue about the level of royalties for weeks on end, but in my experience it always seems to end up in a figure between about 2 and 3%. Maybe others will look at these numbers and be amazed, but every single discussion I have been involved with (for the concepts that were licensed and those for which discussions occurred but nothing resulted) led to a number in this range. Finally, as an academic you will most definitely need the patience of a saint! Like us, industrialists have many demands on their time and there will be occasions when they suddenly go silent for days or even weeks at a time and then return demanding data within 24 hrs. Just heave a sigh, it is all part of the rich tapestry of transferring technology.

b) Licensing (an academic's view of the industrial perspective)

My strong advice to industrialists is to try and get involved as soon as possible. Academics are notorious for feeling that they know what is best and they are perfectly capable, all too often, of developing something that is scientifically excellent but impractical from a business perspective. The sooner that you get involved, the more you can influence the direction of the work and ensure that it develops into something relevant. Though, of course, this does mean that the money will have to flow sooner too. Having said this, it is important to listen to what the academics say. They are the experts (or should be) on what the technology really can deliver. I have come across situations where the industrialists are keen to limit the work to minor tweaks to their existing technology rather than see the delivery of something really revolutionary, but more risky. There is clearly a balance to be achieved here. Also, when suggesting that academics are listened to, don't be swayed by an impressive title, it is the quality of the ideas and the extent of the background that should determine how much you listen, not whether they are a full Professor with an international reputation or a very junior Assistant Professor. What the latter lacks in terms of experience is often more than compensated for by a more open mind. In dealing with the academic, please do remember that they have a day job too, away from their research. They also have to teach students, grade coursework and exams and wade through piles of administration. Please accept that there are times in the academic year when they will simply be swamped. Unless they are very inexperienced, however, they should be able to alert you to when these times are well in advance.

Probably my greatest piece of advice, though, is to be genuinely honest and open with the academics. Get an NDA signed and then deal openly with them. There have been many, many occasions in my career where I knew that the information I was being provided with was incomplete or even, sometimes, deliberately wrong. It is very tempting to believe that all academics leak information, but it actually happens very rarely. Whilst even that might be considered too often, if the work really is so utterly confidential, it should not be happening in a university environment. Research students and postdocs do talk to each other during the course of their work and unless you have the funding to create your own dedicated laboratory (which does happen), then you have to accept that really confidential work should probably not be done in a university. Finally, do expect to need the patience of a saint and don't begrudge a reasonable royalty if the technology works.

Spin out

As you will have seen from the above case studies, I have only had one experience of creating a spin out company and that was twenty years ago now. Many things have changed over the last two decades, one of which is that many more universities have had much more experience of this and have learnt, often the hard way, what to do and what not to do. Although not from personal experience, my major piece of advice to the potentially interested academic is to avoid competing with the industrial partners that helped develop the technology. I watched a fellow academic at another university in the UK do this not very long ago; the industrialists who had helped to develop the technology were really quite bitter and devoted considerable time to ensuring that other companies didn't fund future research with the individual concerned.

With respect to running the business once formed, it is probably best to find an expert to do this, preferably one with very significant experience and a good track record. They are not too difficult to find and most universities will organise this for you (in fact, many won't actually allow you to run your own spin out, the success rates are too low when academics are in charge). It is far better to have a role as an academic consultant, to be on hand whenever needed. Even with this role, you should expect to have to devote lots of time to the business; it really does gobble up time – and, depending on how the workload model works in your Department or School, you can find that academic colleagues resent the time you spend if they feel that you are getting a lighter teaching or admin load whilst undertaking work likely to put money in your pocket. Spin outs also consume a lot of money and every time you need more from business angels or other sources, your share decreases. Nevertheless, if you get it right and the business is successful then it could make you rich (eventually), unlike licensing.

CONCLUSIONS

So, what are the overall conclusions? Certainly, it isn't easy transferring technology from university to industry, the limited number of successes are testimony to this. You need:

- To keep the technology as simple as possible, every complexity added increases the costs;
- To avoid introducing a new product AND a new process at the same time;
- A sound understanding of the technology involved, if you don't have this you WILL be found out sooner or later;
- To see things from the industrial perspective most of the time. Academic work is generally intellectually stimulating, industrial work is focused on making money. There is quite a difference.
- To consider final manufacture from the beginning. This is most important; the last thing that anybody needs is to get halfway through the development and realise that a major step has to be completely redeveloped because the existing approach is simply not economically viable.
- Talented researcher(s) who have an engineering bias. Good researchers will make the whole process both more likely to succeed and more enjoyable. They can take such a load off your shoulders.
- Mutual trust and openness between the company and academic. This is probably the hardest thing to develop; it rarely appears overnight. The academic has to believe that they are not going to find themselves being cheated; the company has to trust that the academic really understands the complexities and limitations brought on by commercialisation.

- Multiple funding sources and luck to tie them together (or very deep industrial pockets). It is so difficult to get continuous funding, just when you need it, from Government-based sources.

Having provided all of the advice above, I still have one more piece to impart. Remember to have fun, it is a lovely feeling when something works and you see the product of your work developments being commercialised!

REFERENCES
1. Y. Yin, J.G.P. Binner, S.J. Marshall and T.E. Cross, Multi-phase ceramics by infiltration processing, in Ceramics: Charting the Future, Ed. P. Vincenzini, *Advances in Science and Technology*, Techna srl, Italy, 2171-2178 (1995).
2. Y. Yin, J.G.P. Binner, T.E. Cross and S.J. Marshall, The oxidation behaviour of carbon fibres, *J. Mat. Sci.* **29** 2250-2254 (1994).
3. A.C. Young, O.O. Omatete, M.A. Janey and P.A. Menchhofer, Gel casting of alumina, *J. Am. Ceram. Soc.* **74** [3] 612-618 (1991).
4. J.G.P. Binner, Foam Ceramics: Part I. Applications and processing routes, *International Ceramics* **1** 76-80 (1998).
5. J.G.P. Binner, Foam Ceramics: Part II. A case study for the development of a new process, *International Ceramics* **2** 69-71 (1998).
6. H. Chang, J. Binner, R.L. Higginson, P. Myers, P. Webb and G. King, Preparation and characterization of ceramic-faced metal-ceramic interpenetrating composites for impact applications, *J. Mat. Sci.* **46** [15] 5237-5244 (2011).
7. J. Liu, J. Binner and R.L. Higginson, Dry sliding wear behaviour of co-continuous ceramic foam / aluminium alloy interpenetrating composites produced by pressureless infiltration, *Wear* **276–277** 94–104 (2012).
8. I. Santacruz, K. Annapoorani and J.G.P. Binner, Preparation of high solids content nano zirconia suspensions, *J. Am. Ceram. Soc.* **91** [2] 398-405 (2008).
9. J. Binner, B. Vaidhyanathan, A. Paul, K. Annapoorani and B. Raghupathy, Compositional effects in nanostructured yttria partially stabilised zirconia, *Int. J. Appl. Ceram. Techn.* **8** [4] 766-782 (2011).
10. S. Eckhard and M. Nebelung: *Proc. 9th Int. Symp. on Agglomeration & 4th Int. Granulation Workshop 2009*, Sheffield, UK. pp. 251-253 (2009).
11. B. Raghupathy and J.G.P. Binner, Spray freeze drying of YSZ nanopowder, *J. Nanoparticle Res.* 14:921 (2012).
12. J. Binner, K. Annapoorani, A. Paul, I. Santacruz and B. Vaidhyanathan, Dense nanostructured zirconia by two stage conventional/hybrid microwave sintering, *J. Eur. Ceram. Soc.* **28** 973-977 (2008).
13. J.G.P. Binner and T.E. Cross, Use of the inverse temperature profile in microwave processing of advanced ceramics, *Microwave Processing of Materials III*, MRS Proceeding **269**, Eds. R.L. Beatty, W.H. Sutton and M.F. Iskander, 357-362 (1992).
14. L.A. Timms, W. Westby, C. Prentice, D. Jaglin, R.A. Shatwell and J.G.P. Binner, Reducing chemical vapour infiltration time for ceramic matrix composites, *J Microscopy* **201** [2] 316-323 (2001).
15. B. Vaidhyanathan, K. Annapoorani, J.G.P. Binner and R. Raghavendra, Microwave assisted large scale sintering of multilayer electroceramic devices, *Ceram. Sci. Eng. Proc.*, **30** [8] 11-18 (2009).
16. I.A.H. Al-Dawery, J. Binner, G. Tari, P.R. Jackson, W.R. Murphy and M. Kearns, Rotary moulding of ceramic hollow ware, *J. Eur. Ceram. Soc.* **29** 887-891 (2009).

17. J.G.P. Binner, A.M. McDermott, Y. Yin, R.M. Sambrook and B. Vaidhyanathan, In-situ coagulation moulding: a new route for high quality, net shape ceramics, *Ceram Int* **32** [1] 29-35 (2006).
18. J.G.P. Binner and A.M. McDermott, In-situ acidification of electrosterically dispersed alumina suspensions: effect of coagulant structure, *J Mat Sci.* **41** 1893-1903 (2006).
19. J. Binner and B. Vaidhayanathan, When should microwaves be used in the processing of ceramics, in *Advances in Ceramic Materials*, ed. P Xiao and B Ralph, *Materials Science Forum*, Trans Tech Publications Ltd, Switzerland, **606**, 51-59 (2009).
20. F.C.R. Wroe, Scaling up the microwave firing of ceramics, *Ceramic Transactions* **36** 449-458 (1993).
21. H.S. Shulman, M.L. Fall and P. Strickland, Ceramic processing using microwave assist technology, *Am. Ceram. Soc. Bull.* **87** [3] 34-36 (2008).

ACKNOWLEDGEMENT

The author thanks all his research team over the years for their outstanding work and the various funding bodies and industry for supporting his work, without which, none of it would have been possible.

Author Index